Mathematical Programming and the Numerical Solution of Linear Equations

Modern Analytic *and* Computational Methods *in* Science *and* Mathematics

A GROUP OF MONOGRAPHS AND ADVANCED TEXTBOOKS

Richard Bellman, EDITOR
University of Southern California

Published

1. R. E. Bellman, R. E. Kalaba, and Marcia C. Prestrud, Invariant Imbedding and Radiative Transfer in Slabs of Finite Thickness, 1963

2. R. E. Bellman, Harriet H. Kagiwada, R. E. Kalaba, and Marcia C. Prestrud, Invariant Imbedding and Time-Dependent Transport Processes, 1964

3. R. E. Bellman and R. E. Kalaba, Quasilinearization and Nonlinear Boundary-Value Problems, 1965

4. R. E. Bellman, R. E. Kalaba, and Jo Ann Lockett, Numerical Inversion of the Laplace Transform: Applications to Biology, Economics, Engineering, and Physics, 1966

5. G. Mikhlin and K. L. Smolitskiy, Approximate Methods for Solution of Differential and Integral Equations, 1967

6. R. N. Adams and E. D. Denman, Wave Propagation and Turbulent Media, 1966

7. R. L. Stratonovich, Conditional Markov Processes and Their Application to the Theory of Optimal Control, 1968

8. A. G. Ivakhnenko and V. G. Lapa, Cybernetics and Forecasting Techniques, 1967

9. G. A. Chebotarev, Analytical and Numerical Methods of Celestial Mechanics, 1967

10. S. F. Feshchenko, N. I. Shkil', and L. D. Nikolenko, Asymptopic Methods in the Theory of Linear Differential Equations, 1967

11. A. G. Butkovskiy, Distributed Control Systems, 1969

12. R. E. Larson, State Increment Dynamic Programming, 1968

13. J. Kowalik and M. R. Osborne, Methods for Unconstrained Optimization Problems, 1968

14. S. J. Yakowitz, Mathematics of Adaptive Control Processes, 1969

15. S. K. Srinivasan, Stochastic Theory and Cascade Processes, 1969

16. D. U. von Rosenberg, Methods for the Numerical Solution of Partial Differential Equations, 1969

17. R. B. Banerji, Theory of Problem Solving; An Approach to Artificial Intelligence, 1969

18. R. Lattès and J.-L. Lions, The Method of Quasi-Reversibility; Applications to Partial Differential Equations. Translated from the French edition and edited by Richard Bellman, 1969

19. D. G. B. Edelen, Nonlocal Variations and Local Invariance of Fields, 1969

20. J. R. Radbill and G. A. McCue, Quasilinearization and Nonlinear Problems in Fluid and Orbital Mechanics, 1970

21. W. Squire, Integration for Engineers and Scientists, 1970

22. T. Parthasarathy and T. E. S. Raghavan, Some Topics in Two-Person Games, 1971

23. T. Hacker, Flight Stability and Con trol, 1970

24. D. H. Jacobson and D. Q. Mayne, Differential Dynamic Programming, 1970

25. H. Mine and S. Osaki, Markovian Decision Processes, 1970

26. W. Sierpinski, 250 Problems in Elementary Number Theory, 1970

27. E. D. Denman, Coupled Modes in Plasmas, Elastic Media, and Parametric Amplifiers, 1970

28. F. H. Northover, Applied Diffraction Theory, 1971

29. G. A. Phillipson, Identification of Distributed Systems, 1971

30. D. H. Moore, Heaviside Operational Calculus; An Elementary Foundation, 1971

31. S. M. Roberts and J. S. Shipman, Two-Point Boundary Value Problems; Shooting Methods, 1971

32. V. F. Demyanov and A. M. Rubinov, Approximate Methods in Optimization Problems, 1970

33. S. K. Srinivasan and R. Vasudevan, Introduction to Random Differential Equations and Their Applications, 1971

34. C. J. Mode, Multitype Branching Processes; Theory and Applications, 1971

35. R. Tomovic and M. Vukobratovic, General Sensitivity Theory, 1971

36. J. G. Krzyz, Problems in Complete Variable Theory, 1972

37. W. T. Tutte, Introduction to the Theory of Matroids, 1971

38. B. W. Rust and W. R. Burrus, Mathematical Programming and the Numerical Solution of Linear Equations, 1972

In Preparation

39. J. O. Mingle, The Invariant Imbedding Theory of Nuclear Transport

40. H. M. Lieberstein, Mathematical Physiology

Mathematical Programming and the Numerical Solution of Linear Equations

Bert W. Rust
Tennecomp, Inc.
Oak Ridge, Tennessee

Walter R. Burrus
Tennecomp, Inc.
Oak Ridge, Tennessee

American Elsevier Publishing Company, Inc.
NEW YORK · 1972

AMERICAN ELSEVIER PUBLISHING COMPANY, INC.
52 Vanderbilt Avenue, New York, N.Y. 10017

ELSEVIER PUBLISHING COMPANY
335 Jan Van Galenstraat, P.O. Box 211
Amsterdam, The Netherlands

International Standard Book Number 0-444-00119-0

Library of Congress Card Number 70-173472

AMS 1970 Subject Classifications
15A06, 45L10, 62F25, 65F30, 90C50

Printed in the United States of America

Contents

Preface

This book is concerned with the solutions to the linear vector, matrix equation

$$\mathbf{Ax} = \mathbf{y} + \mathbf{e},$$

where $\mathbf{x}$ is an n-vector of unknowns, $\mathbf{A}$ is a known $m \times n$ matrix, and $\mathbf{y}$ is a known m-vector estimate of the right-hand side which contains some error or uncertainty $\mathbf{e}$. We shall consider both problems in which the vector $\mathbf{e}$ is known with absolute certainty to lie in some bounded set, and problems in which $\mathbf{e}$ is a statistical uncertainty which can lie anywhere in m-space with some nonzero probability. In the latter case, however, we shall restrict ourselves to problems in which $\mathbf{e}$ is drawn from a joint Normal distribution with completely known variance. We shall also include the case of the Fredholm integral equation

$$\int A(t,s)x(s)ds = y(t) + e(t)$$

in this class of problems as the limiting case when $m, n \to \infty$.

The solution of linear systems of equations is an ancient art; the extensions to cover the statistical character of the right-hand side were initially studied by Gauss. In this book, we focus on the case where the system is poorly conditioned. As far as we know, there is no area of modern applied mathematics in which so many misconceptions exist and so much futile effort is spent as in solving linear systems which are extremely poorly conditioned, singular, or undetermined with $n > m$.

The approach of the more academic statisticians to many problems of this type is to avoid them or refuse to admit that a confidence interval for a singular set of equations has any meaning. But applied workers have reported to innumerous and ingenious techniques which are essentially *ad hoc* or heuristic. Although there are hundreds of papers in the mathematical literature concerning these problems, there is a great schism between the theoretical understanding of the results and the practical application. Thus there is, amongst the many practitioners, a lively and emotional debate about the relative subjective merits of various *ad hoc* devices, and a general despair that things can be put on a rational basis.

Our purpose in writing this book was to show that useful solutions can be obtained to undetermined problems and that valid statistical statements may be made about the results. Furthermore, these rigorous methods have sufficient power that they frequently yield more information about the solution than the heuristically motivated *ad hoc* methods. We wish that this book could have been definitive, but we have raised more questions than we have answered. We hope, however, that we have replaced the older fuzzy questions with newer, sharper ones.

Our approach in writing this book was to make it sufficiently rigorous to be acceptable to theoretical statisticians. Thus, the first two chapters treat only classical statistics and recast it slightly in a form which will be useful later. Essentially, only one new idea is introduced (in Chapter 4)–that of using *a priori* knowledge about the solution to allow statistical statements about the solution. This new idea is introduced in two ways. Chapter 3 gives a Gedanken chapter (an "imaginary thought experiment" description) which, in a pedestrian way, tries to give some applied perspective into the motivations for the use of *ad hoc* knowledge. This chapter can be skipped with no loss in our argument; and in fact, the theoretical statistician will have a hard time accommodating to the jargon which is freely used there. The fourth chapter introduces the key *a priori* constraint idea in a concise mathematical form. Any objection to the overall approach must rest here. The remainder of the book is a straightforward development of the consequences of classical statistics with this one modification.

The authors will judge the book to be minimally successful if it is accepted as interesting (and hopefully useful) by applied workers and at the same time is not criticized too sharply by theoreticians. We will judge the book to be highly successful if it succeeds in stimulating further work on some of the key problem areas we have pin-pointed. Namely:

- Sharpening the limits on constrained confidence intervals given in Chapter 4.
- Development of algorithms for solution of the special quadratic programming problem (which is central to the estimation problem given in Chapter 5) which requires less computer time.
- Removal of some of the restrictive assumptions which we have made for exposition.

We are indebted to the Defense Atomic Support Agency for support in the writing of this book and for courage in supporting the objective to write a book which attempts to reconcile two very deeply entrenched biases. We wish to express our sincere appreciation to Miss Vicky Gall for her tireless, accurate and patient preparation of the manuscript and the index.

Oak Ridge, Tennessee

BERT W. RUST
WALTER R. BURRUS

CHAPTER 1

Ill-Conditioned Linear Systems

1.1 Introduction

Probably the best (or at least the most widely acclaimed) classical method for solving nonsingular linear systems on a computer is Gaussian elimination with pivoting and iterative refinement. This procedure is discussed in great detail in Forsythe and Moler's book *Computer Solution of Linear Algebraic Systems* [1], and Wilkinson has given extensive treatments of the error analysis of the method in a paper in 1961 [2] and in his book *Rounding Errors in Algebraic Processes* [3]. The method is fast, and if the matrix of the system is not too ill-conditioned, it gives correctly rounded approximations to the true solution.

In this monograph, we will be concerned with solving ill-conditioned systems. By an ill-conditioned system, we mean a system

$$\mathbf{A}\mathbf{x} = \mathbf{b}$$

such that small relative changes in the matrix $\mathbf{A}$ or in the right-hand-side vector $\mathbf{b}$ make very large relative changes in the solution vector $\mathbf{x}$. By itself, an ill-conditioned system is very distressing; because there will inevitably be errors in the matrix $\mathbf{A}$ and in the vector $\mathbf{b}$. Such errors will arise from the initial shortening of the input numbers when the problem is read into a finite-accuracy computer. There might also be inherent uncertainties in the elements of $\mathbf{A}$ and $\mathbf{b}$ if, for example, the quantities are measured data. Whatever their source, the presence of these errors makes it impossible to obtain a meaningful solution by simply applying one of the classical methods to the system by itself. Moreover, if all we know is the system itself, there is no nonclassical method that will give a meaningful solution. In this regard, it is good to keep in mind the statement by Lanczos that "a lack of information cannot be remedied by any mathematical trickery" [4]. But in many applications some other information is known about the solution. It is often easy to get some a priori information about the solution by considering the original source of the system; this monograph is concerned with how to incorporate that a priori information into the method of solution so that a meaningful answer can be obtained.

An estimate of the solution to a linear system is much more valuable if it is accompanied by an estimate of the error. The most often-used approach in the error analysis of linear systems is *inverse error analysis,* which has been developed mainly by Wilkinson. Another approach is *interval analysis,* as developed by Hansen [5, 6] and Moore [7]. In the latter method, the solution consists of an interval for each of the elements of **x**, and the algorithm is designed so that the interval is guaranteed to contain the true value of that element.

We will also treat the solution of ill-conditioned systems as a problem of interval estimation and incorporate the a priori information about the solution into the problem as constraints. The resulting constrained estimation problem can then be reduced to a problem in mathematical programming that can be solved to give confidence intervals for the true solution of the system.

Since we do not seek a unique single number for each element of the solution vector **x**, we do not have to confine our attention to systems with nonsingular matrices. In fact, the system matrix need not even be square, and we will consider rectangular matrices that may or may not have full rank. In cases where the matrix has less than full rank, there is no unique solution and the confidence interval that we will obtain for a given element of **x** will be a confidence interval for the set of all possible values of that element that are consistent with the a priori information about the solution.

Since an ill-conditioned system is one in which the solution vector **x** is a highly unstable function of the elements of the matrix **A** and of the right-hand-side vector **b**, it is tempting to regard problems that generate such systems as being improperly formulated or ill conceived. Although this is undoubtedly true in many cases, it is not always true; ill-conditioned systems arise quite naturally in many practical problems. One very important example that we will discuss at great length is the numerical solution of integral equations of the first kind. These equations are not often discussed in the literature on integral equations, but they are very important in many branches of the physical sciences.

1.2 The Fredholm Integral Equation of the First Kind

The Fredholm integral equation of the first kind has the form

$$\int_a^b K(t,s)\,x(s)\,ds = y(t), \qquad a \leqslant t \leqslant b, \tag{1.2-1}$$

where $K(t, s)$ and $y(t)$ are known functions, a and b are known constants, and $x(s)$ is an unknown function. The functions $K(t, s)$ and $y(t)$ are usually assumed to be continuous or at least bounded and piecewise continuous. The theoretical treatment of this equation is complicated, and most texts on integral equations do not discuss it at very great length. It will not have a solution for every function

$y(t)$, and when it is solvable there may be no unique solution. Even if a unique solution exists, it may be highly unstable in the sense that small changes in the function $y(t)$ produce very large changes in the solution $x(s)$.

The function $K(t, s)$ is called the *kernel* of the integral equation. Any nontrivial function $\phi(t)$ satisfying

$$\int_a^b K(t,s)\,\phi(s)\,ds = \lambda\phi(t), \qquad a \leqslant t \leqslant b, \tag{1.2-2}$$

for some constant λ is called an *eigenfunction* of the kernel. The constant λ is the *eigenvalue* corresponding to that eigenfunction. Several linearly independent eigenfunctions may correspond to the same value λ, but we will speak of the set of eigenvalues as if they are all distinct with one eigenvalue λ_n corresponding to each distinct linearly independent eigenfunction $\phi_n(t)$. Clearly, if some of the eigenvalues are equal to zero, that is, if

$$\int_a^b K(t,s)\,\phi_n(s)\,ds = 0 \tag{1.2-3}$$

for one or more of the eigenfunctions, then there will not be a unique solution to Eq. (1.2-1), because any linear combination of eigenfunctions satisfying Eq. (1.2-3) can be added to any solution of Eq. (1.2-1) and the result is also a solution. The kernel $K(t, s)$ is said to be *closed* if it does not have any zero eigenvalues; that is, if there are no nontrivial functions $\phi_n(s)$ that satisfy Eq. (1.2-3). Thus Eq. (1.2-1) can have a unique solution only if $K(t, s)$ is closed.

The kernel $K(t, s)$ is said to be *separable* if it is expressible in the form

$$K(t,s) = \sum_{n=1}^{N} f_n(t)\,g_n(s) \tag{1.2-4}$$

where N is a finite integer and the functions $f_1(t), f_2(t), \ldots, f_N(t)$ are linearly independent on the interval $[a, b]$. Clearly, if the kernel is separable, then Eq. (1.2-1) has a solution only if the right-hand side $y(t)$ is a linear combination of the functions $f_1(t), f_2(t), \ldots, f_N(t)$. Even then the solution is not unique, for there will be an infinity of functions $h_\nu(s)$ that are othogonal to the subspace spanned by the functions $g_1(s), g_2(s), \ldots, g_N(s)$. The complex conjugate $h_\nu^*(s)$ of any such $h_\nu(s)$ is an eigenfunction corresponding to a zero eigenvalue, because

$$\int_a^b K(t,s)\,h_\nu^*(s)\,ds = \sum_{n=1}^{N} f_n(t)\left[\int_a^b g_n(s)\,h_\nu^*(s)\,ds\right]$$

and each of the integrals in the sum on the right vanishes.

A real kernel $K(t, s)$ is said to be *symmetric* if $K(t, s) = K(s, t)$, and a complex kernel is symmetric if $K(t, s) = K^*(s, t)$, where $K^*(s, t)$ is the complex conjugate of $K(s, t)$. The eigenvalues of a symmetric kernel are all real, and any two eigenfunctions corresponding to unequal eigenvalues are orthogonal; that is,

$$\int_a^b \phi_m^*(s)\,\phi_n(s)\,ds = 0$$

if $\lambda_m \neq \lambda_n$. Moreover, the set of eigenfunctions corresponding to all the eigenvalues that are equal to some common value can be replaced by an orthogonal set containing the same number of eigenfunctions. Hence, we can always pick an orthonormal set of eigenfunctions for a symmetric kernel. Let $\lambda_1, \lambda_2, \ldots, \lambda_n, \ldots$ be the eigenvalues, and suppose that

$$|\lambda_1| \geqslant |\lambda_2| \geqslant |\lambda_3| \geqslant \ldots \geqslant |\lambda_n| \geqslant \ldots.$$

Let $\phi_1(t), \phi_2(t), \ldots, \phi_n(t), \ldots$ be the corresponding eigenfunctions and suppose that $\{\phi_n(t)\}$ is an orthonormal set; that is, suppose that

$$\int_a^b \phi_m^*(s)\,\phi_n(s)\,ds = \begin{cases} 0, & m \neq n, \\ 1, & m = n. \end{cases}$$

Then the kernel can be expressed as

$$K(t, s) = \sum_{n=1}^{\infty} \lambda_n\,\phi_n(t)\,\phi_n^*(s), \tag{1.2-5}$$

provided that the (Fourier) series on the right converges uniformly.

If $K(t, s)$ is symmetric but has only a finite number N of nonzero eigenvalues, then $K(t, s)$ is separable and Eq. (1.2-5) reduces to

$$K(t, s) = \sum_{n=1}^{N} \lambda_n\,\phi_n(t)\,\phi_n^*(s). \tag{1.2-6}$$

In this case, the integral equation (1.2-1) is solvable only if the right-hand side $y(t)$ is expressible as a linear combination of the eigenfunctions $\phi_1(t), \phi_2(t), \ldots, \phi_N(t)$. If $y(t)$ is given by

$$y(t) = \alpha_1\,\phi_1(t) + \alpha_2\,\phi_2(t) + \ldots + \alpha_N\,\phi_N(t) \tag{1.2-7}$$

where $\alpha_1, \alpha_2, \ldots, \alpha_N$ are constants (not all equal to zero), then one solution to Eq. (1.2-1) is

$$x(s) = \frac{\alpha_1}{\lambda_1}\,\phi_1(s) + \frac{\alpha_2}{\lambda_2}\,\phi_2(s) + \ldots + \frac{\alpha_N}{\lambda_N}\,\phi_N(s). \tag{1.2-8}$$

But this is not a unique solution, because any linear combination of the eigenfunctions $\phi_{N+1}(t), \phi_{N+2}(t), \phi_{N+3}(t), \ldots$ can be added to it and the result is also a solution.

If $K(t, s)$ is symmetric and closed, then the set of orthonormal eigenfunctions $\{\phi_n(t)\}$ is a complete set, and the right-hand side $y(t)$ can be expanded in a Fourier series in terms of the $\phi_n(t)$. If the Fourier expansion of $y(t)$ is

$$y(t) = \sum_{n=1}^{\infty} \alpha_n \phi_n(t), \tag{1.2-9}$$

where the constants α_n are given by

$$\alpha_n = \int_a^b y(s) \phi_n^*(s)\, ds, \tag{1.2-10}$$

then (1.2-1) has a unique solution if and only if the series

$$\sum_{n=1}^{\infty} \frac{|\alpha_n|^2}{|\lambda_n|^2}$$

converges. If this series does converge, then the solution is

$$x(s) = \sum_{n=1}^{\infty} \frac{\alpha_n}{\lambda_n} \phi_n(s). \tag{1.2-11}$$

For the general case when $K(t, s)$ is not symmetric, two *associated symmetric kernels* can be defined by

$$L(t, s) = \int_a^b K(u, t) K^*(u, s)\, du, \tag{1.2-12}$$

$$R(t, s) = \int_a^b K(t, u) K^*(s, u)\, du. \tag{1.2-13}$$

Both $L(t, s)$ and $R(t, s)$ are symmetric nonnegative functions; they both have real nonnegative eigenvalues and orthonormal sets of eigenfunctions, and any nonzero eigenvalue of one of them is also an eigenvalue of the other. The positive square roots of these common (positive) eigenvalues are called *singular values* of the kernel $K(t, s)$. Let these common eigenvalues (squares of the singular values) be

$$\sigma_1{}^2 \geqslant \sigma_2{}^2 \geqslant \ldots \geqslant \sigma_n{}^2 \geqslant \ldots,$$

and suppose that the corresponding eigenfunctions for $L(t, s)$ are $\eta_1(t), \eta_2(t), \ldots, \eta_n(t), \ldots$ and the corresponding eigenfunctions for $R(t, s)$ are $\psi_1(t), \psi_2(t), \ldots, \psi_n(t), \ldots$. Since $L(t, s)$ and $R(t, s)$ are symmetric, each of the two sets $\eta_n(t)$ and $\psi_n(t)$ can be chosen to be orthonormal on the interval $[a, b]$. Neither of the two sets is necessarily complete, since either or both of the associated kernels may have some zero eigenvalues. The two sets are related by

$$\int_a^b K(t,s)\,\eta_n(s)\,ds = \sigma_n \psi_n(t), \tag{1.2-14}$$

$$\int_a^b K^*(s,t)\,\psi_n(s)\,ds = \sigma_n \eta_n(t); \tag{1.2-15}$$

each of the pairs $\eta_n(t), \psi_n(t)$ is called a pair of *adjoint singular functions* belonging to the singular value σ_n. In many cases, the kernel may be written

$$K(t,s) = \sum_{n=1}^{\infty} \sigma_n \psi_n(t)\,\eta_n^*(s); \tag{1.2-16}$$

in many other cases,

$$K(t,s) = \underset{N\to\infty}{\text{l.i.m.}} \sum_{n=1}^{N} \sigma_n \psi_n(t)\,\eta_n^*(s), \tag{1.2-17}$$

where l.i.m. denotes convergence in the mean.

If the kernel $K(t, s)$ is separable, then the series in Eq. (1.2-16) will have only a finite number N of terms, and the integral equation (1.2-1) will be solvable only if the right-hand side $y(t)$ is a linear combination of $\psi_1(t), \psi_2(t), \ldots, \psi_N(t)$. In the case where the associated kernel $R(t, s)$ is closed, the set of eigenfunctions $\psi_n(t)$ is a complete orthonormal set and $y(t)$ can be expressed as a Fourier series in terms of the $\psi_n(t)$:

$$y(t) = \sum_{n=1}^{\infty} \beta_n \psi_n(t). \tag{1.2-18}$$

The integral equation (1.2-1) will then have a solution if and only if the series

$$\sum_{n=1}^{\infty} \frac{|\beta_n|^2}{\sigma_n^2}$$

converges. If the associated kernel $L(t, s)$ is also closed, the solution is unique and is given by

$$x(s) = \sum_{n=1}^{\infty} \frac{\beta_n}{\sigma_n} \eta_n(s).$$

The discussion of the integral equation (1.2-1) that we have given here is very incomplete, and many important details have been omitted. More rigorous treatments can be found in Part I of Pogorzelski's *Integral Equations and Their Applications* [8] and in Smithies' *Integral Equations* [9]. In addition, Smithies' 1937 paper [10] in the *Proceedings of the London Mathematical Society* is a very good source on eigenvalues and singular values.

1.3 Applications of the Fredholm Equation

The integral equation (1.2-1)

$$\int_a^b K(t,s)\,x(s)\,ds = y(t)$$

arises in many branches of the physical sciences. It becomes especially important in the experimental sciences whenever physical data are measured by indirect sensing devices [11, 12, 13]; in this context it is sometimes called the resolution-correction equation. If the errors of measurement are taken into account, the equation is usually written

$$\hat{y}(t) = \int_a^b K(t,s)\,x(s)\,ds + \hat{\epsilon}(t), \qquad a \leqslant t \leqslant b, \tag{1.3-1}$$

where $x(s)$ is the physical quantity whose value is desired, $y(t)$ is the physical quantity that can actually be measured, and $\hat{\epsilon}(t)$ is a stochastic function that represents the random error of measurement and whose expectation is

$$E[\hat{\epsilon}(t)] = 0, \qquad a \leqslant t \leqslant b.$$

The stochastic function $\hat{y}(t)$ represents the measured value of $y(t)$, and

$$E[\hat{y}(t)] = y(t), \qquad a \leqslant t \leqslant b.$$

The kernel $K(t, s)$ is the *response function* for the instrument, and it contains the information on how the quantity $x(s)$ interacts with the instrument to produce the measured quantity $y(t)$.

The ideal response for a measuring instrument would be

$$K(t,s) = \delta(t-s), \qquad a \leqslant t, s \leqslant b,$$

for then

$$y(t) = \int_a^b K(t,s)\,x(s)\,ds = \int_a^b \delta(t-s)\,x(s)\,ds = x(t), \qquad a \leqslant t \leqslant b,$$

and the quantity actually measured would be the quantity whose measurement is sought. Of course, this is not obtainable, and the next best situation would be a response function like the one illustrated in Fig. 1.1. For each fixed value of t, $K(t, s)$ is a sharply peaked function of s with the peak centered at $s = t$ and dropping off rapidly to zero on both sides of the peak. Figure 1.2 illustrates the interaction of such a response function with a hypothetical function $x(s)$, which has four peaks superimposed upon a continuum. The first peak is broad compared to the width of the response function, the second two peaks are of comparable width

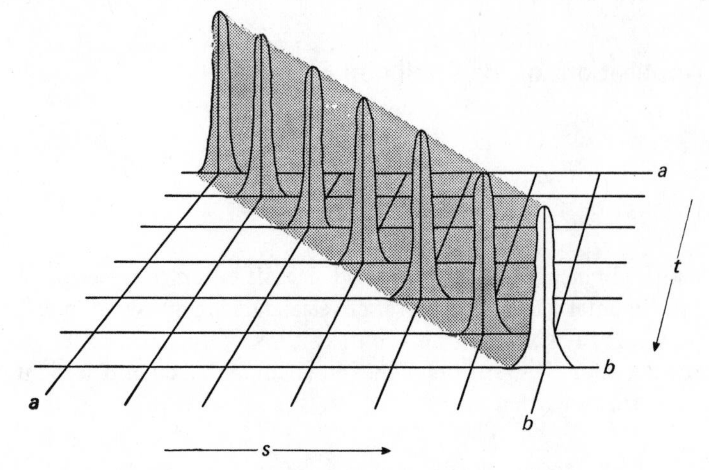

Figure 1.1. A hypothetical response function $K(t, s)$.

to the response function, and the third peak is much sharper than the response function. The measured curve $y(t)$ has the same general shape as the $x(s)$ distribution, but because the peaks in the response function are not infinitely sharp,

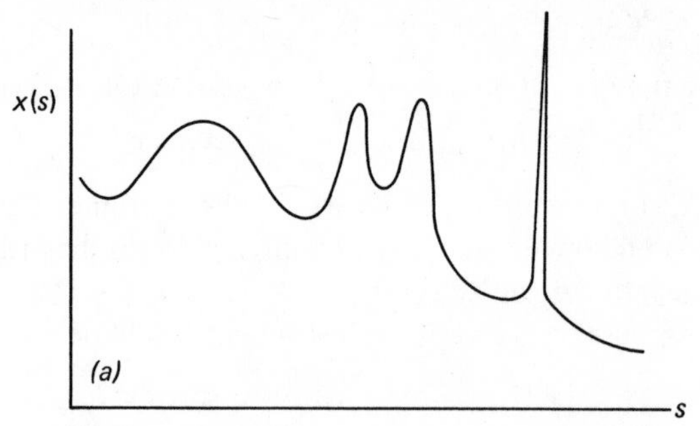

(a) Function whose measurement is desired.

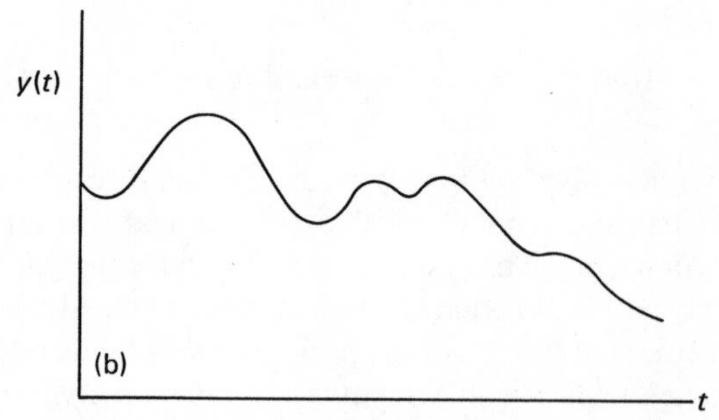

(b) Actual measurement obtained.

Figure 1.2.

the peaks in $x(s)$ are smoothed, and some of the detail of the spectrum is lost. In particular, the second and third peaks are almost unresolved, and the height of the fourth peak has been greatly reduced. This loss of resolution becomes more and more noticeable as the peaks in $K(t, s)$ become wider relative to those in the function $x(s)$. In actual practice, the situation is even worse, because many measuring instruments have response functions like the one shown in Fig. 1.3. For

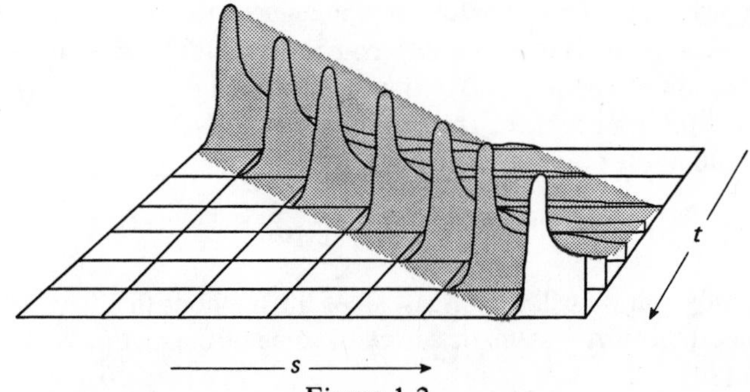

Figure 1.3.

each value of t, $K(t, s)$ has a peak at $s = t$, but for $s \geqslant t$, the function tails off very slowly to zero. Also, for some instruments the peaks are not located exactly on the line $s = t$, and frequently the peak width varies with s and t. Some instruments have response functions that do not even have distinct peaks.

The presence of the uncertainty $\hat{\epsilon}(t)$ in Eq. (1.3-1) aggravates the normal problems of nonuniqueness and instability of the solution. Even if the solution of

$$y(t) = \int_a^b K(t, s)\, x(s)\, ds$$

is unique, there may be many functions $g(s)$ that satisfy

$$\left| \int_a^b K(t, s)\, g(s)\, ds \right| < |\hat{\epsilon}(t)|,$$

and for any such function the sum $x(s) + g(s)$ satisfies the integral equation within the limits of the experimental uncertainty. It can be shown, for example, that if $K(t, s)$ is integrable, then

$$\int_a^b K(t, s) \sin(ns)\, ds \to 0 \qquad \text{as } n \to \infty,$$

so that an arbitrarily large amount of a high-frequency sinusoidal component could be added to any solution, and the result would still be consistent with the measured values when the experimental error is taken into account. Thus, for any measuring instrument that is characterized by Eq. (1.3-1), the function $x(s)$ may have *invisible components* that the instrument cannot see. These invisible components can arise from two sources. If the response function $K(t, s)$ is not closed, then the instrument cannot see any component in $x(s)$ that is a multiple of an eigenfunction corresponding to a zero eigenvalue. The other source of invisible components is the statistical error in $y(t)$, which, when it is combined with the loss of resolution discussed in the preceding paragraph, makes it impossible to detect high-frequency oscillatory components or to distinguish between the many such components that would be compatible with the measured data.

The problem of obtaining $x(s)$ from the measured distribution $y(t)$ is sometimes called the *unfolding* problem. This terminology stems from the German usage of *Faltung* (folding) for the integral

$$\int_a^b K(t,s)\,x(s)\,ds.$$

In this monograph we will not discuss all of the methods that have been used to attack the problem. A systematic survey of some of the methods was given by Monahan [14].

In a given experimental situation, either or both of s and t may be discrete variables. One particular problem that we will consider in detail has s as a continuous variable and t as a discrete variable. Such problems often arise in experimental nuclear physics when a multichannel pulse-height analyzer or a gamma-ray scintillation spectrometer is used to measure energy spectra. Figure 1.4 is a sketch of the response function for a gramma-ray scintillation spectrometer. In this case, the variable s becomes the continuous variable E, which is the energy of the photons being measured, and the variable t becomes the discrete variable i, which is the channel number for the channels or bins in which the photon counts are accumulated. The kernel or response function consists of a collection of single-valued functions $K(i, E) = K_i(E), i = 1, 2, \ldots, m$, one for each channel. For any given energy $E, K_i(E)$ is just the probability that a photon of that energy would produce a count in the ith channel. The quantity whose measurement is desired is the spectral distribution $g(E)$ of the incident photons and the quantity that can actually be measured is the number of counts C_i in the ith channel. The two are related by

$$C_i = \int_0^\infty K_i(E)\,g(E)\,dE, \qquad i = 1, \ldots, m,$$

and since the experimenter can usually determine some upper bound E_u on the energy of the incident photons, this relationship can be written

$$C_i = \int_0^{E_u} K_i(E)\,g(E)\,dE, \qquad i = 1, \ldots, m. \tag{1.3-2}$$

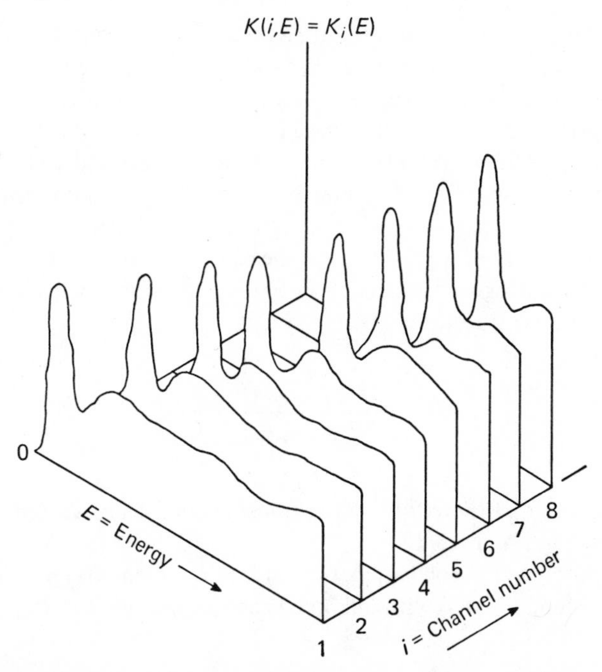

Figure 1.4. Sketch of the response function for a gamma-ray scintillation spectrometer.

Taking the random measuring error into account gives

$$C_i = \int_0^{E_u} K_i(E) g(E)\, dE + \hat{\epsilon}_i, \qquad i = 1, \ldots, \mathrm{m}. \tag{1.3-3}$$

We have already remarked that the presence of this uncertainty rules out any hope that a unique solution will be defined by the measured data.

1.4 Numerical Solutions of Fredholm Equations

An excellent survey of some of the classical methods for solving the Fredholm equation

$$\int_a^b K(t,s)\, x(s)\, ds = y(t), \qquad a \leqslant t \leqslant b, \tag{1.4-1}$$

is given in Chapter 1.4 of Trumpler and Weaver's *Statistical Astronomy* [15]. It is possible to obtain direct solutions only in certain special cases. Eddington [16] has given a series solution derived from a Taylor expansion, which is applicable if

$a=-\infty, b=+\infty$, and $K(t, s)$ has the form $K(t-s)$. The solution series will be semi-convergent if the values of the kernel $K(t-s)$ rapidly become negligible outside the peaks and if $x(s)$ and $y(t)$ are relatively smooth in comparison to $K(t-s)$. Lane, Morehouse, and Phillips [17] have given a method for solving this same problem when the limits of integration are finite. The method requires that $K(s-t)$ and $y(t)$ be expanded in a Fourier series on $[a, b]$, and it gives the solution $x(s)$ as a Fourier series whose coefficients are determined by the coefficients of the series for K and y. In the case where the limits of integration are infinite, the analogous approach uses Fourier transforms with the solution given by the inverse transform

$$x(s)=\frac{1}{2\pi}\int_{-\infty}^{\infty}\frac{F_y(u)}{F_K(u)}e^{-ius}\,du,$$

where $F_y(u)$ and $F_K(u)$ are the Fourier transforms of y and K, respectively (see Trumpler and Weaver [15]).

Another approach to solving Eq. (1.4-1) is to represent $x(s)$ by some general interpolating function $p(s)$ with unknown coefficients, and to carry out the integration

$$I(t)=\int_a^b K(t,s)p(s)\,ds$$

using an interpolating function for $K(t, s)$ also when it is not available as an exact analytical expression. The result of the integration $I(t)$ will be an interpolating function for $y(t)$ containing the unknown constants from the function $p(s)$. These unknown coefficients are then determined so that $I(t)$ gives the best possible fit to $y(t)$. If a good fit can be obtained, the values of the coefficients thus determined are substituted into $p(s)$, and the resulting function is used as an approximation to $x(s)$.

In most of the situations in which a solution of (1.4-1) is required, the previously discussed methods are not applicable. The usual approach is to obtain a numerical solution by replacing the integral equation with an equivalent system of linear algebraic equations whose solution will be a pointwise approximation to $x(s)$. Formally, the linear integral equation (1.4-1) is equivalent to an infinite set of linear algebraic equations with an infinite number of unknowns. In practice the problem is discretized, and the integral equation is replaced by a finite linear system. The discretization may be carried out in two steps. The first step is to replace the continuous variable t with a finite set of mesh points $t_1, t_2, \ldots, t_m$,

$$a \leqslant t_1 < t_2 < \ldots < t_m \leqslant b$$

and to write

$$\begin{aligned}
y(t_1) &= \int_a^b K(t_1, s)\, x(s)\, ds, \\
y(t_2) &= \int_a^b K(t_2, s)\, x(s)\, ds, \\
&\cdots\cdots\cdots\cdots \\
y(t_m) &= \int_a^b K(t_m, s)\, x(s)\, ds.
\end{aligned} \tag{1.4-2}$$

There is no discretization error in this step, and any solution of Eq. (1.4-1) will satisfy this system of equations exactly. There may, however, be many functions $x(s)$ that satisfy this system of equations but are not solutions to Eq. (1.4-1).

The second step of the discretization process is the replacement of the continuous variable s with a finite set of mesh points $s_1, s_2, \ldots, s_n$ such that

$$a \leqslant s_1 < s_2 < \ldots < s_n \leqslant b.$$

To carry out this step, each of the integrals in Eq. (1.4-2) must be replaced by a numerical quadrature formula,

$$\begin{aligned}
y(t_1) &= \int_a^b K(t_1, s)\, x(s)\, ds \cong \sum_{j=1}^{n} w_j K(t_1, s_j)\, x(s_j), \\
y(t_2) &= \int_a^b K(t_2, s)\, x(s)\, ds \cong \sum_{j=1}^{n} w_j K(t_2, s_j)\, x(s_j), \\
&\cdots\cdots\cdots\cdots\cdots\cdots \\
y(t_m) &= \int_a^b K(t_m, s)\, x(s)\, ds \cong \sum_{j=1}^{n} w_j K(t_m, s_j)\, x(s_j),
\end{aligned} \tag{1.4-3}$$

where $w_1, w_2, \ldots, w_n$ are the weighting coefficients for the quadrature formula used. The use of the quadrature sum introduces a discretization error that varies inversely with the number of points in the s mesh.

If the vectors **x** and **y** are defined by

$$\mathbf{x} = \begin{bmatrix} x_1 \\ x_2 \\ \cdot \\ \cdot \\ \cdot \\ x_n \end{bmatrix} = \begin{bmatrix} x(s_1) \\ x(s_2) \\ \cdot \\ \cdot \\ \cdot \\ x(s_n) \end{bmatrix}, \qquad \mathbf{y} = \begin{bmatrix} y_1 \\ y_2 \\ \cdot \\ \cdot \\ \cdot \\ y_m \end{bmatrix} = \begin{bmatrix} y(t_1) \\ y(t_2) \\ \cdot \\ \cdot \\ \cdot \\ y(t_m) \end{bmatrix}, \tag{1.4-4}$$

and if the $m \times n$ matrix $\mathbf{K}$ is defined by

$$\mathbf{K} = [K_{ij}] = \begin{bmatrix} w_1 K(t_1, s_1) & w_2 K(t_1, s_2) & \dots & w_n K(t_1, s_n) \\ w_1 K(t_2, s_1) & w_2 K(t_2, s_2) & \dots & w_n K(t_2, s_n) \\ \dots\dots\dots & \dots\dots\dots & \dots & \dots\dots\dots \\ w_1 K(t_m, s_1) & w_2 K(t_m, s_2) & \dots & w_n K(t_m, s_n) \end{bmatrix}, \tag{1.4-5}$$

then Eq. (1.4-2) can be written

$$\mathbf{y} = \mathbf{K}\mathbf{x} \tag{1.4-6}$$

when the discretization error is ignored. This vector matrix equation represents m linear equations in n unknowns. The usual procedure in classical numerical methods is to choose m equal to n. The resulting $n \times n$ matrix $\mathbf{K}$ will, on account of the discretization error, usually be nonsingular, in spite of the fact that the system of equations in (1.4-2) that it replaces may not have a unique solution. The linear system (1.4-6) is also poorly conditioned. This poor conditioning is not caused by the discretization error but is inherent in the integral equation (1.4-1) itself. We saw in Section 1.3 that if there is a small uncertainty $\epsilon(t)$ in the right-hand side $y(t)$, then there are many widely different functions $x(s)$ that satisfy (1.4-1) within the limits of the uncertainty. Clearly, then, a very small change in $y(t)$ can make a very large change in $x(s)$, and this instability will be reflected in the linear system (1.4-6). In fact, choosing n larger and larger in order to reduce the discretization error will make the system (1.4-6) more, rather than less, ill conditioned, because the larger we choose n, the more closely does (1.4-6) approximate the integral equation (1.4-1). The linear system (1.4-6) can be solved by any of the standard methods, such as Gaussian elimination, but it should be clear from the foregoing remarks that the solution thus obtained may be a very poor pointwise approximation to the solution $x(s)$.

Baker *et al.* [18] have given a numerical method, based on Eq. (1.2-11), for solving Fredholm equations with symmetric kernels. Although (1.2-11) gives the solution explicitly in terms of the eigenvalues and eigenfunctions of $K(t, s)$ it would not be practical to compute the solution directly from this equation, because the problem of obtaining the complete eigensystem $\{\lambda_n, \phi_n(s)\}$ and the Fourier expansion of $y(t)$ in terms of the eigenfunctions $\{\phi_n(t)\}$ is just as formidable as the original integral equation. But if we discretize the problem, choosing $m=n$ and choosing the t mesh points to be the same as the s mesh points used in quadrature formula, that is,

$$t_i = s_i, \qquad i = 1, 2, \ldots, n,$$

and if we write the resulting linear system in the form

$$\mathbf{K}'\mathbf{D}\mathbf{x} = \mathbf{y} \tag{1.4-7}$$

where

$$\mathbf{K}' = \begin{bmatrix} K(t_1,s_1) & K(t_1,s_2) & \ldots & K(t_1,s_n) \\ K(t_2,s_1) & K(t_2,s_2) & \ldots & K(t_2,s_n) \\ \cdot & \cdot & & \cdot \\ \cdot & \cdot & & \cdot \\ \cdot & \cdot & & \cdot \\ K(t_n,s_1) & K(t_n,s_2) & \ldots & K(t_n,s_n) \end{bmatrix} \tag{1.4-8}$$

and

$$\mathbf{D} = \operatorname{diag}(w_1, w_2, \ldots, w_n), \tag{1.4-9}$$

that is, a diagonal matrix with the weighting coefficients of the quandrature formula as the diagonal elements, then the matrix $\mathbf{K}'$ will be symmetric. Let the eigenvalues of $\mathbf{K}'$ be $\lambda_1, \lambda_2, \ldots, \lambda_n$ and the corresponding eigenvectors be $\boldsymbol{\phi}_1$, $\boldsymbol{\phi}_2, \ldots, \boldsymbol{\phi}_{n_j}$. The set $\{\boldsymbol{\phi}_n\}$ can be chosen to be orthonormal, and the matrix $\mathbf{K}'$ can be written

$$\mathbf{K}' = \lambda_1 \boldsymbol{\phi}_1 \boldsymbol{\phi}_1^* + \lambda_2 \boldsymbol{\phi}_2 \boldsymbol{\phi}_2^* + \ldots + \lambda_n \boldsymbol{\phi}_n \boldsymbol{\phi}_n^* \tag{1.4-10}$$

where the row vector $\boldsymbol{\phi}_j^*$ is the conjugate transpose of the column vector $\boldsymbol{\phi}_j$. The right-hand side of (1.4-7) can be written

$$\mathbf{y} = \alpha_1 \boldsymbol{\phi}_1 + \alpha_2 \boldsymbol{\phi}_2 + \ldots + \alpha_n \boldsymbol{\phi}_n \tag{1.4-11}$$

where the α_j are given by

$$\alpha_j = \boldsymbol{\phi}_j^* \mathbf{y}, \qquad j = 1, \ldots, n. \tag{1.4-12}$$

The solution vector $\mathbf{x}$ is given by

$$\mathbf{x} = \mathbf{D}^{-1}\left[\frac{\alpha_1}{\lambda_1}\boldsymbol{\phi}_1 + \frac{\alpha_2}{\lambda_2}\boldsymbol{\phi}_2 + \ldots + \frac{\alpha_n}{\lambda_n}\boldsymbol{\phi}_n\right]. \tag{1.4-13}$$

All the remarks in the preceding paragraph about the ill-conditioning of the system (1.4-6) are applicable to this method also.

1.5 Examples

In this section we will give some examples of Fredholm equations of the first kind and of the use of classical numerical techniques for solving them. The examples that we will present have been given previously by other authors, who also suggested methods for improving the poor solutions that the classical methods yield.

The first example is the integral equation

$$\int_0^1 (t-s)^2\, x(s)\, ds = \frac{t^2}{2} - \frac{2t}{3} + \frac{1}{4}, \qquad 0 \leqslant t \leqslant 1, \tag{1.5-1}$$

which has been discussed by Bellman, Kalaba, and Lockett [19]. In this example, the kernel $K(t, s) = (t-s)^2$ is symmetric and hence has all real eigenvalues and orthogonal eigenfunctions. The only three nonzero eigenvalues are

$$\lambda_0 = -\frac{1}{6}, \qquad \lambda_1 = \frac{5 - 3\sqrt{5}}{60}, \qquad \lambda_2 = \frac{5 + 3\sqrt{5}}{60},$$

and the corresponding eigenfunctions are

$$\phi_0(t) = 2t - 1,$$

$$\phi_1(t) = t^2 - t + \frac{5 - \sqrt{5}}{20},$$

$$\phi_2(t) = t^2 + t + \frac{5 + \sqrt{5}}{20}.$$

There is an infinity of zero eigenvalues, and the functions

$$\phi_p(t) = L_p(2t - 1), \qquad p = 3, 4, 5, \ldots,$$

where $L_p(x)$ is the pth-order Legendre polynomial, comprise an orthogonal set of eigenfunctions corresponding to the zero eigenvalues. One solution of Eq. (1.5-1) is

$$x(s) = s, \tag{1.5-2}$$

and this solution is unique in the subspace spanned by the eigenfunctions $\phi_0(t)$, $\phi_1(t)$, and $\phi_2(t)$; but any linear combination of $\phi_3(t), \phi_4(t), \phi_5(t), \ldots$ can be added to this solution, and the resulting sum is also a solution of (1.5-1).

If the problem is discretized using the points $s_i = t_i = 0.1(i)$, $i = 0, 1, 2, \ldots,$ 10, and the Simpson rule quadrature formula, the resulting 11 x 11 linear system is

$$\mathbf{Kx} = \mathbf{y}$$

where

$$y_i = \frac{t_i^2}{2} - \frac{2t_i}{3} + \frac{1}{4}, \qquad i = 1, 2, \ldots, 11,$$

$$K_{ij} = \frac{1}{10} w_j (t_i - s_j)^2, \qquad i, j = 1, 2, \ldots, 11,$$

and the w_j are the weighting coefficients for Simpson's rule; that is

$$(w_1, w_2, w_3, w_4, \ldots, w_9, w_{10}, w_{11}) = \frac{1}{3}(1, 4, 2, 4, \ldots, 2, 4, 1).$$

This example is rather special because the kernel is a quadratic function of s; so for $x(s) = s$, the Simpson's rule formula gives the exact value of the integral without any discretization error. Therefore, if the components of the matrix **K** are computed exactly (without any rounding errors), **K** will be a singular matrix of rank 3. In actual practice, when the matrix **K** is formed in a computer, there will be rounding errors and errors due to the initial shortening of the input data; the resulting matrix **K** will be nonsingular but very poorly conditioned. There will also be initial shortening errors in the components of the vector **y**. Figure 1.5 illustrates the results of solving the linear system by matrix inversion. The solution

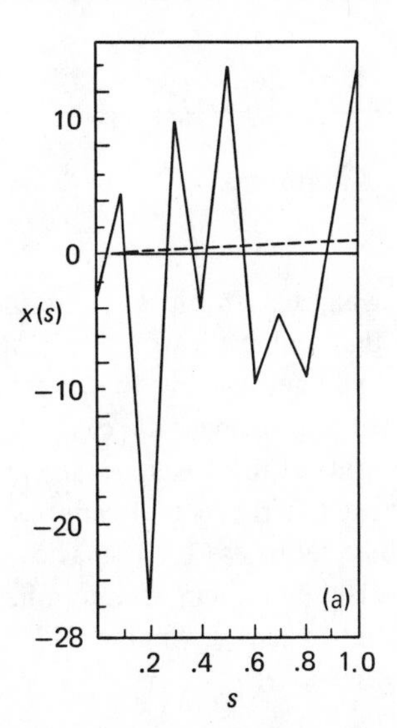

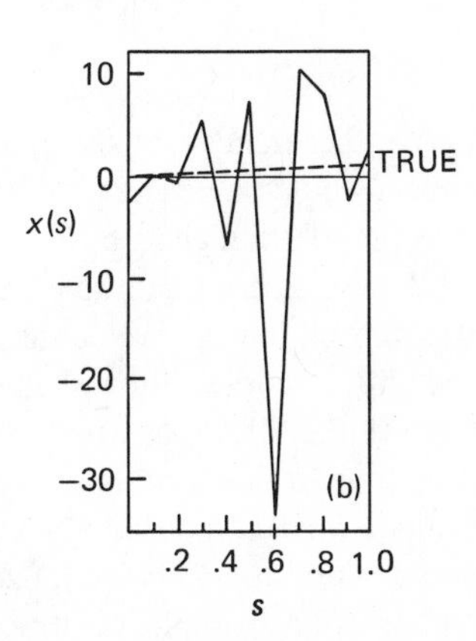

Figure 1.5. (a) Solution by matrix inversion obtained by Bellman, *et. al.* with right-hand side accurate to eight significant figures.

(b) Solution by matrix inversion obtained by present authors, with right-hand side accurate to eight decimal digits.

is a widely oscillating function and changes quite radically with small changes in the right side **y**. The differences between the two solutions shown arise from differences in the ninth significant figure of the matrix **K** and the right-hand side and differences in the matrix inversion routines used to obtain the solutions.

Although (1.5-1) does not have a unique solution, the solution (1.5-2) is unique in the subspace corresponding to the nonzero eigenvalues. There is clearly no hope of finding this unique solution by simple matrix inversion. We might instead try the eigensystem approach and compute a solution by Eq. (1.4-13). Computing the eigenvalues of matrix **K**, using computer arithmetic, would give fairly accurately the three nonzero eigenvalues but would give small numerical

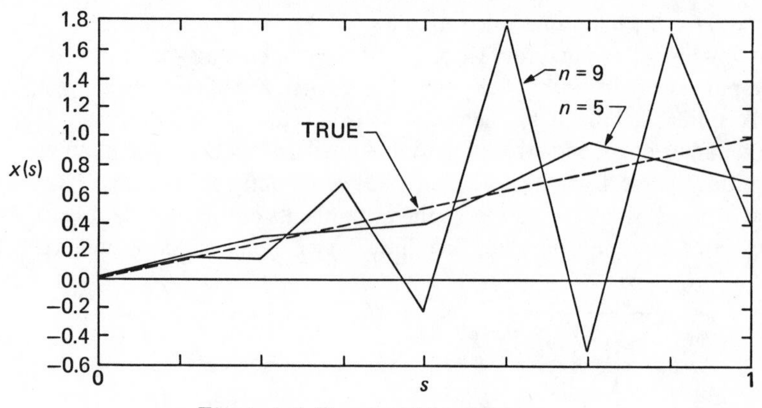

Figure 1.6. Trapezoidal quadrature.

values, rather than exact zeros, for the remaining eight eigenvalues. These small numbers would be divisors in Eq. (1.4-13), and the final result would be just as bad as the solution given by matrix inversion. If we knew a priori that there are only three nonzero eigenvalues, or if we suspected that this was the case after computing all the eigenvalues and noticing that eight of them were small compared with the other three, then we could compute a fairly accurate approximation to the solution by using only the first three terms of (1.4-13). The difficulty with this technique in the general case is the problem of determining whether a small computed eigenvalue should be set to zero or whether it is an approximation to an actual nonzero but small eigenvalue.

As a second example, consider the integral equation

$$\int_0^1 (t^2 + s^2)^{1/2} x(s)\, ds = \tfrac{1}{3}[(1 + t^2)^{3/2} - t^3], \qquad 0 \leqslant t \leqslant 1, \qquad (1.5\text{-}3)$$

which has been discussed by Fox and Goodwin [20]. The kernel $K(t, s) = (t^2 + s^2)^{1/2}$ is closed, and the unique solution is

$$x(s) = s, \qquad 0 \leqslant s \leqslant 1. \tag{1.5-4}$$

Fox and Goodwin discretized the problem using three different quadrature formulas, the trapezoidal rule, Simpson's rule, and Gaussian quadrature. For each of these methods, they carried out the discretization, using several different mesh spacings for s, and solved the resulting linear system. The problem was again solved by the present authors using discretizations of $n = 5, n = 9$, and $n = 13$. In each case, m was chosen to be equal to n, and the resulting linear system was solved by Gaussian elimination. Figures 1.6 and 1.7 show the results

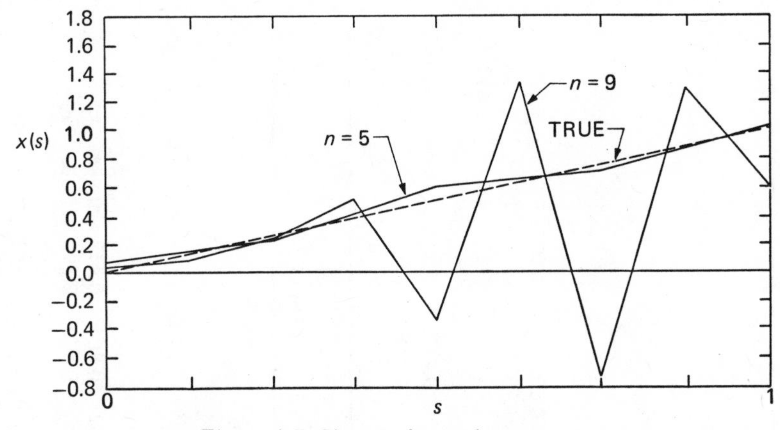

Figure 1.7. Simpson's quadrature.

for the two more widely separated mesh spacings for the trapezoidal and Simpson quadratures, respectively. Tables 1.1 and 1.2 show the complete results for all three quadratures.

The powerful Gaussian quadrature gives the best approximation to the solution for $n = 9$, but also yields the most wildly oscillating solution when the kernel is most accurately approximated for $n = 13$. These tables clearly show that smaller and smaller intervals first make the solutions better by removing some of the errors in the approximation to the kernel, but then cause wild oscillations when the kernel is so well approximated that the poor conditioning of the system overwhelms the solution.

As a third example, consider

$$\int_{-6}^{6} K(t, s)\, x(s)\, ds = y(t), \qquad |t| \leqslant 6, \tag{1.5-5}$$

TABLE 1.1

n	$x(s) = s$ (true solution) $f_{\text{Trap}}(s)$ (solution using trapezoidal quadrature) $f_{\text{Simp}}(s)$ (solution using Simpson's quadrature)												
	0.000	0.250	0.500	0.750	1.000								
5	0.029	0.283	0.399	0.957	0.688								
	0.044	0.212	0.599	0.718	1.032								
	0.000	0.125	0.250	0.375	0.500	0.625	0.750	0.875	1.000				
9	0.014	0.146	0.157	0.678	– 0.222	1.793	– 0.481	1.716	0.408				
	0.021	0.109	0.236	0.509	– 0.333	1.344	– 0.722	1.287	0.612				
	0.000	0.083	0.167	0.250	0.333	0.417	0.500	0.583	0.667	0.750	0.833	0.917	1.000
13	0.009	0.095	0.131	0.175	1.850	– 8.758	33.99	– 80.75	134.5	– 146.5	104.4	– 40.95	15.66
	0.015	0.065	0.387	– 0.761	14.47	– 33.55	227.0	– 262.9	849.2	– 460.9	646.8	– 129.8	93.96

TABLE 1.2

n	$x(s) = s$ (true solution) $f_{\text{Gaus}}(s)$ (solution using Gaussian quadrature)											
5	0.046	0.230	0.500	0.769	0.953							
	0.047	0.228	0.503	0.764	0.957							
9	0.015	0.081	0.193	0.337	0.500	0.662	0.806	0.918	0.984			
	0.016	0.081	0.193	0.336	0.502	0.655	0.820	0.897	1.003			
12	0.009	0.479	0.115	0.206	0.316	0.437	0.562	0.683	0.793	0.884	0.953	0.990
	0.036	– 213.2	0.200	0.071	– 0.345	37.98	130.1	– 247.6	539.6	– 970.7	1352.1	– 1252.0

where

$$K(t,s) = \begin{cases} 1 + \cos\dfrac{\pi(s-t)}{3}, & |s-t| \leqslant 6, \quad |t| \leqslant 6, \\ 0, & |s-t| \geqslant 3, \quad |t| \leqslant 6, \end{cases} \tag{1.5-6}$$

and

$$y(t) = \begin{cases} (6-|t|)\left[1 + \tfrac{1}{2}\cos\dfrac{\pi t}{3}\right]\dfrac{9}{2\pi} + \sin\dfrac{\pi|t|}{3}, & |t| \leqslant 6, \\ 0, & |t| \geqslant 6. \end{cases} \tag{1.5-7}$$

These two functions are illustrated in Figs. 1.8 and 1.9. This problem has been discussed by several authors, but was originally given by Phillips [21]. The kernel

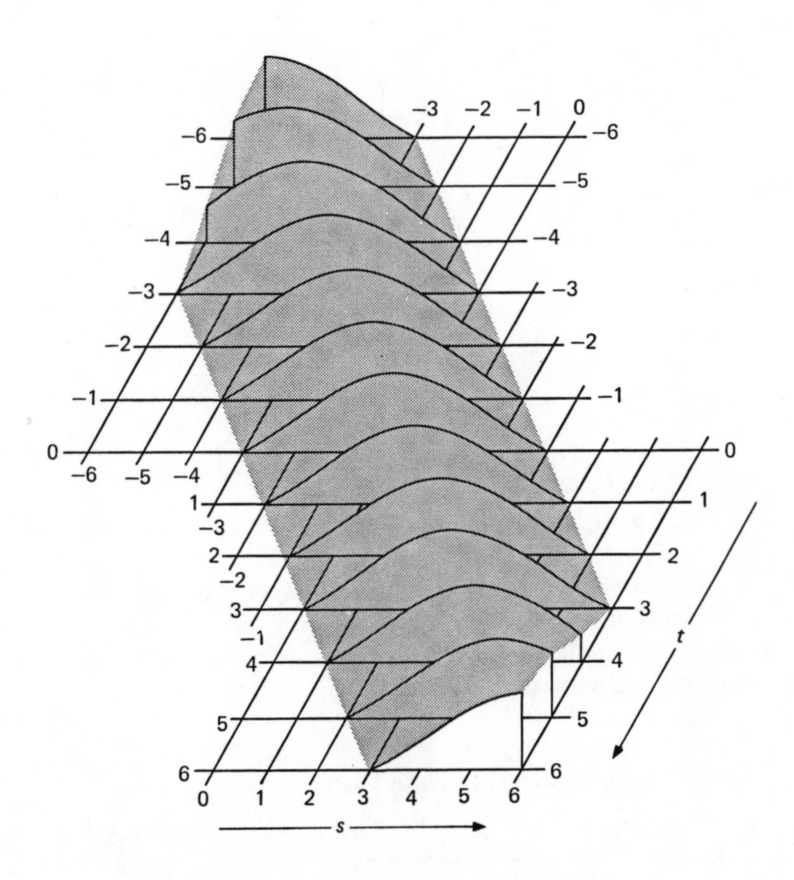

Figure 1.8. Kernel function for the example of Phillips.

is symmetric (and closed), and the solution is

$$x(s) = \begin{cases} 1 + \cos \dfrac{\pi s}{3}, & |s| \leqslant 3, \\ 0, & |s| \geqslant 3. \end{cases} \tag{1.5-8}$$

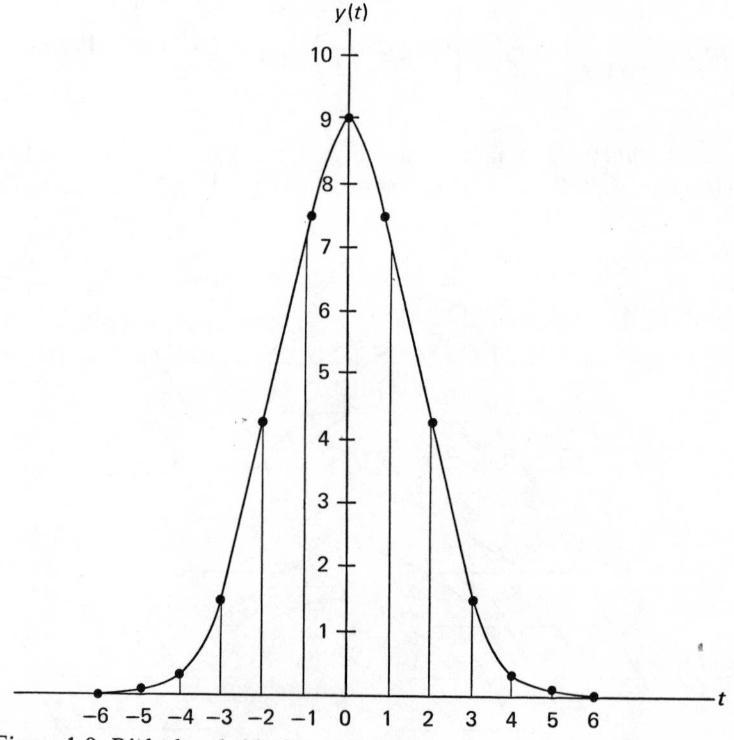

Figure 1.9. Right-hand side function $y(t)$ for the example of Phillips.

Figure 1.10 shows the true solution and the results of discretizing the problem using Simpson's rule and solving the resulting set of linear equations by Gaussian elimination. These results again show that a more closely approximated kernel does not yield a more accurate solution.

1.6 Smoothing and Regularizing Techniques

Since the Fredholm equation of the first kind is so important for applied work in the physical sciences, there have been many workers who have sought to avoid the ill-conditioning problem by requiring their solution algorithms to produce

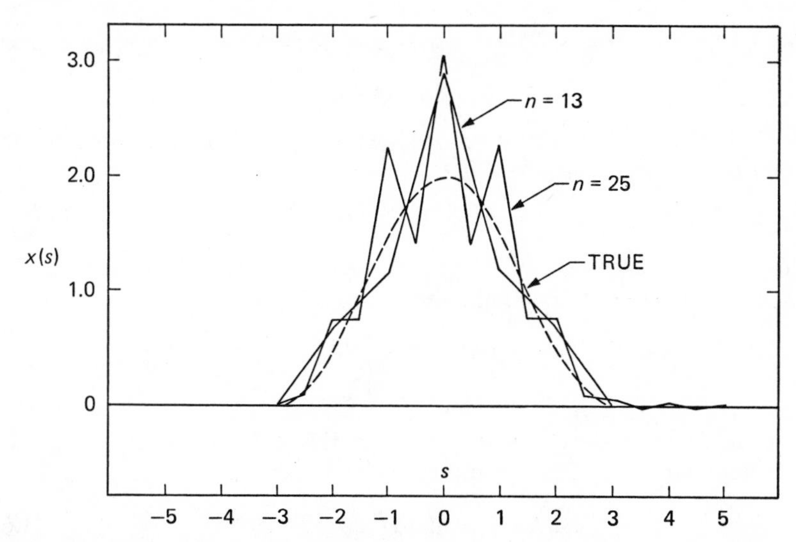

Figure 1.10. Solutions of Phillips' problem using Simpson's rule for two different mesh spacings.

only solutions that also satisfy various a priori side constraints that are suggested by known physical properties of the solution. One important class of methods of this type are the methods of Phillips [21] and Twomey [22], who treated the integral equation in the form

$$\int_a^b K(t,s)\,x(s)\,ds = y(t) + \epsilon(t), \qquad a \leqslant t \leqslant b, \tag{1.6-1}$$

where the "error" function $\epsilon(t)$ is arbitrary except for some restriction on its magnitude. The family of allowed functions $\epsilon(t)$ defines a family X of allowed solutions $x(s)$, but most of these solutions will not be physically meaningful.

Phillips assumed that the desired true solution is reasonably smooth and tried to find the function $x(s) \in X$ that is smoothest in some sense. He assumed that the required $x(s)$ has a piecewise continuous second derivative $x''(s)$, and he chose as the smoothness criterion the condition that

$$\int_a^b [x''_{\text{sm}}(s)]^2\,ds = \min_{x \in X} \int_a^b [x''(s)]^2\,ds. \tag{1.6-2}$$

If the continuous variables s and t are each replaced by an equally spaced mesh,

$$a = s_1 < s_2 < s_3 < \ldots < s_n = b,$$
$$a = t_1 < t_2 < t_3 < \ldots t_n = b,$$

with $s_i = t_i$, $i = 1, 2, 3, \ldots, n$, then (1.6-1) is replaced by the vector-matrix equation

$$\mathbf{Kx} = \mathbf{y} + \boldsymbol{\epsilon} \tag{1.6-3}$$

by the discretization process described in Section 1.4. Phillips required that the vector ϵ satisfy the constraint

$$\|\boldsymbol{\epsilon}\|^2 = \boldsymbol{\epsilon}^T \boldsymbol{\epsilon} = e^2 \tag{1.6-4}$$

where e^2 is a constant. He used second differences to approximate second derivatives and wrote the smoothness constraint (1.6-2) in the form

$$\sum_{i=1}^{n} (x_{i+1}^{\text{sm}} - 2x_i^{\text{sm}} + x_{i-1}^{\text{sm}})^2 = \min_{\mathbf{x} \in \mathbf{X}^*} \sum_{i=1}^{n} (x_{i+1} - 2x_i + x_{i-1})^2.$$

where $\mathbf{X}^*$ is the set of vectors satsifying (1.6-3) and (1.6-4), and $x_0 = x_{n+1} = 0$. In vector matrix terms, this condition can be written

$$\mathbf{x}_{\text{sm}}^T \mathbf{C} \mathbf{x}_{\text{sm}} = \min_{\mathbf{x} \in \mathbf{X}^*} \mathbf{x}^T \mathbf{C} \mathbf{x} \tag{1.6-5}$$

where $\mathbf{C}$ is the $n \times n$ matrix defined (for the case $n = 8$) by

$$\mathbf{C} = \begin{bmatrix} 1 & -2 & 1 & 0 & 0 & 0 & 0 & 0 \\ -2 & 5 & -4 & 1 & 0 & 0 & 0 & 0 \\ 1 & -4 & 6 & -4 & 1 & 0 & 0 & 0 \\ 0 & 1 & -4 & 6 & -4 & 1 & 0 & 0 \\ 0 & 0 & 1 & -4 & 6 & -4 & 1 & 0 \\ 0 & 0 & 0 & 1 & -4 & 6 & -4 & 1 \\ 0 & 0 & 0 & 0 & 1 & -4 & 5 & -2 \\ 0 & 0 & 0 & 0 & 0 & 1 & -2 & 1 \end{bmatrix}$$

In order to obtain the required smooth solution, the method of Phillips seeks to minimize the quadratic form $\mathbf{x}^T \mathbf{C} \mathbf{x}$ subject to the constraints expressed by Eqs. (1.6-3) and (1.6-4). The result obtained by Phillips was

$$\mathbf{x}_{\text{sm}} = [\mathbf{K} + \gamma (\mathbf{K}^{-1})^T \mathbf{C}]^{-1} \mathbf{y} \tag{1.6-6}$$

where γ is an undefined Lagrange multiplier whose value controls the amount of smoothing used in defining the solution–the greater the value of γ, the greater the amount of smoothing. Phillips also showed that the vector ϵ is related to γ and $\mathbf{x}_{\text{sm}}$ by

$$\boldsymbol{\epsilon} = -\gamma (\mathbf{K}^{-1})^T \mathbf{C} \mathbf{x}_{\text{sm}}.$$

Thus, the idea of the method is to determine by trial and error the largest value of γ that produces a vector ϵ that satisfies the constraint (1.6-4). The corresponding value of x_{sm} is taken to be the smoothed solution to the problem.

Twomey simplified and generalized the method of Phillips. Using the Phillips smoothing criterion, he showed that the solution can be written

$$\mathbf{x}_{sm} = (\mathbf{K}^T \mathbf{K} + \gamma \mathbf{C})^{-1} \mathbf{K}^T \mathbf{y}. \tag{1.6-7}$$

It is easy to show that this solution is completely equivalent to the Phillips solution (1.6-6), but it is somewhat simpler because it involves only one matrix inversion rather than two. Twomey also showed how to generalize the method in order to use a priori constraints other than the Phillips smoothness constraint. This is accomplished by replacing the matrix **C** by a new matrix that is appropriate for the particular constraint used. As an example, he considered the case where the general shape and order of magnitude of the solution are known a priori. The strategy that he adopted was to construct a trial solution $\mathbf{p} \cong \mathbf{x}$ and then to minimize the sums of the squares of the deviations from the trial solution subject to the constraints (1.6-3) and (1.6-4). That is, he sought to minimize the quantity

$$L = (\mathbf{x} - \mathbf{p})^T (\mathbf{x} - \mathbf{p}) + \gamma^{-1} \boldsymbol{\epsilon}^T \boldsymbol{\epsilon}$$

where $\boldsymbol{\epsilon} = \mathbf{Kx} - \mathbf{y}$ and γ^{-1} is an undetermined Lagrange multiplier. The result that he obtained was

$$\mathbf{x} = (\mathbf{K}^T \mathbf{K} + \gamma \mathbf{I})^{-1} (\mathbf{K}^T \mathbf{y} + \gamma \mathbf{p}), \tag{1.6-8}$$

the matrix **C** having been replaced by the identity matrix **I**. Again, the method is to determine the proper value of γ by trial and error. The foregoing expression is a particular case of the more general expression

$$\mathbf{x} = (\mathbf{K}^T \mathbf{K} + \gamma \mathbf{H})^{-1} (\mathbf{K}^T \mathbf{y} + \gamma \mathbf{p}),$$

which can be used to calculate the solutions corresponding to many different a priori constraints by choosing the appropriate values of **H** and **p**.

Another important method for computing stable solutions to (1.6-1) is the regularization method of A. N. Tichonov [23–25]. Tichonov assumes that there exists a unique solution $\bar{x}(s)$ corresponding to $\epsilon(t) = 0$, and that the function $y(t)$ is such that $\bar{x}(s)$ is a continuous piecewise smooth function. The strategy of the regularization method is based upon the functional

$$M^\alpha[x(s), y(t)] = \int_a^b \left[\int_a^b K(t,s)\,x(s)\,ds - y(t) \right]^2 dt + \alpha \int_a^b [k(s)\,x'(s)^2 + p(s)\,x(s)^2]\,ds \tag{1.6-9}$$

where $k(s)$ and $p(s)$ are fixed positive functions and α is a variable parameter. The first term on the right-hand side of the foregoing equation, that is

$$N[x(s), y(t)] \equiv \int_a^b \left[\int_a^b K(t,s)\,x(s)\,ds - y(t) \right]^2 dt, \tag{1.6-10}$$

is just the quadratic deviation

$$\delta^2 = \int_a^b \epsilon(t)^2\,dt,$$

and the second term

$$\alpha\Omega[x(s)] = \alpha \int_a^b [k(s)\,x'(s)^2 + p(s)\,x(s)^2]\,ds \tag{1.6-11}$$

is called the *regularizing functional.* The method consists essentially of determining the function $x^\alpha(s)$ that minimizes $M^\alpha[x(s), y(t)]$ for a given value of α and using the fact (which was proved by Tichonov) that $x^\alpha(s)$ converges uniformly to $\bar{x}(s)$ as $\alpha \to 0$.

In doing the actual calculations, Eq. (1.6-9) must be discretized. If the mesh spacings on s and t are taken to be Δs and Δt with the mesh points denoted by

$$a \leqslant s_1 < s_2 < \ldots < s_n \leqslant b,$$
$$a \leqslant t_1 < t_2 < \ldots < t_m \leqslant b,$$

and if the derivative $x'(s)$ is approximated by first differences, that is,

$$x'(s_j) \cong \frac{x(s_{j+1}) - x(s_j)}{\Delta s},$$

then Eq. (1.6-9) is approximated by

$$M^\alpha[\mathbf{x}, \mathbf{y}] = \Delta t(\mathbf{Kx} - \mathbf{y})^T(\mathbf{Kx} - \mathbf{y}) + \alpha\mathbf{x}^T\left(\frac{1}{\Delta s}\mathbf{C} + \Delta s\mathbf{P}\right)\mathbf{x} \tag{1.6-12}$$

where $\mathbf{x}, \mathbf{y}$, and $\mathbf{K}$ are the vectors and matrix described in Section 1.4, $\mathbf{C}$ is the matrix (in the case $n = 5$)

$$\mathbf{C} = \begin{bmatrix} k_1 & -k_1 & 0 & 0 & 0 \\ -k_1 & k_1 + k_2 & -k_2 & 0 & 0 \\ 0 & -k_2 & k_2 + k_3 & -k_3 & 0 \\ 0 & 0 & -k_3 & k_3 + k_4 & -k_4 \\ 0 & 0 & 0 & -k_4 & k_4 \end{bmatrix}$$

with $k_j = k(s_j)$, and $\mathbf{P}$ is the diagonal matrix

$$\mathbf{P} = \mathrm{diag}(p_1, p_2, p_3, p_4, p_5)$$

with $p_j = p(s_j)$. Tichonov showed that for every value of $\alpha > 0$, there exists a solution vector $\mathbf{x}^\alpha$ that minimizes $M^\alpha[\mathbf{x}, \mathbf{y}]$, the vector $\mathbf{x}^\alpha$ being the solution of the linear system of equations

$$\Delta t[\mathbf{K}^T \mathbf{K}\mathbf{x} - \mathbf{K}^T \mathbf{y}] + \alpha \left[\frac{1}{\Delta s} \mathbf{C} + \Delta s \mathbf{P} \right] \mathbf{x} = 0.$$

Thus $\mathbf{x}^\alpha$ is given by

$$\mathbf{x}^\alpha = \Delta t \left[\Delta t \mathbf{K}^T \mathbf{K} + \alpha \left(\frac{1}{\Delta s} \mathbf{C} + \Delta s \mathbf{P} \right) \right]^{-1} \mathbf{K}^T \mathbf{y}. \tag{1.6-13}$$

The computational method, then, is to choose, by trial and error, the smallest value of the parameter α that yields a stable solution vector $\mathbf{x}^\alpha$.

The method just described makes no restrictions on the function $k(s)$ and $p(s)$ other than that they be positive valued. Tichonov also showed that the method works if the function $p(s)$ is taken to be identically zero, and he also showed that the method can be generalized to use a regularizing functional of the form

$$\alpha \Omega^{(n)}[x(s)] = \alpha \int_a^b \left\{ \sum_{i=1}^{n+1} k_i(s)[x^{(i)}(s)]^2 \right\} ds$$

where $x^{(i)}(s)$ is the ith derivative of $x(s)$ and the $k_i(s)$ are positive continuous functions. The similarity of all these methods to the methods of Phillips and Twomey is readily apparent.

Another method for solving (1.6-1), which is similar in spirit to the techniques just discussed but is different in the method of solution, is the dynamic programming approach of Bellman, Kalaba, and Lockett [19, 26]. In this approach, the discretized problem (1.6-3) is replaced by a minimization problem of the form

$$L = \min_{\mathbf{x}} \{(\mathbf{Kx} - \mathbf{y})^T (\mathbf{Kx} - \mathbf{y}) + \lambda\phi(\mathbf{x})\}$$

where λ is a variable parameter and $\phi(\mathbf{x})$ is a suitably chosen function of the vector $\mathbf{x}$. Thus, the overall approach is very much like the preceding methods. But the method of solution differs in that dynamic programming is employed to obtain the minimizing solution rather than the classical linear equations approach. The details of this method are excellently described in an earlier monograph in this same series [27]. The authors also show how the technique can be applied as a successive approximations method with the results of one step of the computational algorithm being used to guide the procedure in the next step. This approach stems from their concept of a computational solution as an adaptive control process whose ultimate purpose is the minimization of the overall error.

References

1 George Forsythe and Cleve D. Moler, *Computer Solution of Linear Algebraic Systems,* Prentice-Hall, Englewood Cliffs, New Jersey, 1967.
2 J. H. Wilkinson, Error analysis of direct methods of matrix inversion, *J. Assoc. Comput. Mach.,* 8 (1961), 281–330.
3 J. H. Wilkinson, *Rounding Errors in Algebraic Processes,* Prentice-Hall, Englewood Cliffs, New Jersey, 1963.
4 Cornelius Lanczos, *Linear Differential Operators,* Van Nostrand, Princeton, New Jersey, 1961.
5 Eldon Hansen, Interval arithmetric in matrix computations, part I, *SIAM J. Numer. Anal.* 2 (1965), 308–320.
6 Eldon Hansen and Roberta Smith, Interval arithmetic in matrix computations, part II, *SIAM J. Numer. Anal.,* 4 (1967), 1–9.
7 Ramon E. Moore, *Interval Analysis,* Prentice-Hall, Englewood Cliffs, New Jersey 1966.
8 W. Pogorzelski, *Integral Equations and Their Applications,* Vol. I. Macmillan (Pergamon), New York, and Polish Scientific Publishers, Warsaw, 1966.
9 F. Smithies, *Integral Equations,* Cambridge Univ. Press, London and New York, 1958.
10 F. Smithies, The eigen-values and singular values of integral equations, *Proc. London Math. Soc.* (Ser. 2), **43** (1937), 255–279.
11 Carl Eckart, The correction of continuous spectra for the finite resolution of the spectrometer, *Phys. Rev.* **51** (1937), 735–738.

12 R. N. Bracewell and J. A. Roberts, Aerial smoothing in radio astronomy, *Australian J. Phys.* 7 (1954), 615–640.
13 F. D. Kahn, The correction of observational data for instrumental band width, *Proc. Cambridge Phil. Soc.* **51** (1955), 519–525.
14 J. E. Monahan, Unfolding measured distributions, in *Scintillation Spectroscopy* (Steven M. Shaferoth, ed.), Chapter VIII. Gordon and Breach, New York, 1967.
15 R. J. Trumpler and H. F. Weaver, *Statistical Astronomy,* Dover, New York, 1962.
16 A. S. Eddington, On a formula for correcting statistics for the effects of a known probable error of observation, *Monthly Notices Roy. Astronom. Soc.* **73** (1913), 356–360.
17 R. O. Lane, N. F. Morehouse, and D. L. Phillips, The correction of resonance peaks for experimental resolution, *Nucl. Instr. Methods,* 9 (1960), 87–91.
18 C. T. H. Baker, L. Fox, D. F. Mayers, and K. Wright, Numerical solution of Fredholm integral equations of the first kind, *Computer J.* 7 (1965), 141–148.
19 R. Bellman, R. Kalaba, and J. Lockett, Dynamic programming and ill-conditioned linear systems, *J. Math. Anal. Appl.* **10** (1965), 206–215.
20 L. Fox and E. T. Goodwin, The numerical solution of non-singular linear integral equations, *Phil. Trans. Roy. Soc. London* A **245** (1953), 501–534.
21 David L. Phillips, A technique for the numerical solution of certain integral equations of the first kind, *J. Assoc. Comput. Mach.* **9** (1962), 84–97.
22 S. Twomey, On the numerical solution of Fredholm integral equations of the first kind by the inversion of the linear system produced by quadrature, *J. Assoc. Comput. Mach.* **10** (1963), 97–101.
23 A. N. Tichonov, Solution of incorrectly formulated problems and the regularization method, *Soviet Math. Dokl.* 4 (1963), 1035–1038.
24 A. N. Tichonov, Regularization of incorrectly posed problems, *Soviet Math. Dokl.* 4 (1963), 1624–1627.
25 A. N. Tichonov and V. B. Glasko, The approximate solution of Fredholm integral equations of the first kind, *USSR Comput. Math. and Math. Phys.* 4 (3) (1964), 236–247.
26 R. Bellman, R. Kalaba, and J. Lockett, Dynamic programming and ill-conditioned linear systems–II, *J. Math. Anal. Appl.* **12** (1965), 393–400.
27 R. Bellman, R. Kalaba, and J. Lockett, *Numerical Inversion of the Laplace Transform,* Chapter 5. American Elsevier, New York, 1966.

CHAPTER 2

Linear Estimation

2.1 The Linear Regression Model

We saw in Chapter 1 that many physical measurements are characterized by the integral equation

$$\hat{y}(t) = \int_a^b K(t,s)\, x(s)\, ds + \hat{\epsilon}(t), \qquad a \leqslant t \leqslant b, \tag{2.1-1}$$

where $x(s)$ is the physical quantity whose measure is desired, $K(t, s)$ is the response function of the instrument, and $y(t)$ is the quantity that can actually be measured. The stochastic function $\hat{\epsilon}(t)$ is the random measuring error, and $\hat{y}(t)$ is the actual measured value. We also saw that the integral equation

$$\int_a^b K(t,s)\, x(s)\, ds = y(t)$$

can be discretized, using a numerical quadrature formula for the integration, to yield the vector-matrix equation

$$\mathbf{y} = \mathbf{Kx}$$

where $\mathbf{y}$ is the m-vector

$$\mathbf{y}^T = (y(t_1), y(t_2), \ldots, y(t_m)),$$

$\mathbf{x}$ is the n-vector

$$\mathbf{x}^T = (x(s_1), x(s_2), \ldots, x(s_n)),$$

and $\mathbf{K}$ is the $m \times n$ matrix with elements

$$K_{ij} = w_j K(t_i, s_j), \qquad i = 1, 2, \ldots, m, j = 1, 2, \ldots, n,$$

the w_j being the weighting coefficients of the quadrature formula.

Similarly, if the stochastic vectors $\hat{\mathbf{y}}$ and $\hat{\boldsymbol{\epsilon}}$ are defined by

$$\hat{\mathbf{y}}^T = (\hat{y}(t_1), \hat{y}(t_2), \ldots, \hat{y}(t_m)),$$

$$\hat{\boldsymbol{\epsilon}}^T = (\hat{\epsilon}(t_1), \hat{\epsilon}(t_2), \ldots, \hat{\epsilon}(t_m)),$$

then Eq. (2.1-1) can be discretized to give

$$\hat{\mathbf{y}} = \mathbf{K}\mathbf{x} + \hat{\boldsymbol{\epsilon}}. \tag{2.1-2}$$

Each of the elements of the vector $\hat{\mathbf{y}}$ is a quantity obtained from a measurement and the corresponding element in the vector $\hat{\boldsymbol{\epsilon}}$ is the random error in that measurement. In most experiments the measurement errors have zero expectations, that is,

$$E(\hat{\boldsymbol{\epsilon}}) = \mathbf{0} \qquad \text{or} \qquad E(\hat{\mathbf{y}}) = \mathbf{K}\mathbf{x} \tag{2.1-3}$$

and the variance matrix of the measurements

$$E[(\hat{\mathbf{y}} - \mathbf{K}\mathbf{x})(\hat{\mathbf{y}} - \mathbf{K}\mathbf{x})^T] = E(\hat{\boldsymbol{\epsilon}}\hat{\boldsymbol{\epsilon}}^T) = \mathbf{S} \tag{2.1-4}$$

is symmetric and positive definite. Of course, Eq. (2.1-2) does not exactly correspond to Eq. (2.1-1) because of the discretization error that arises when the integral is replaced by a quadrature sum. But this discretization error can be made arbitrarily small, and in particular it can be made negligible in comparison to the experimental error by picking a large enough value for n.

The problem posed by Eqs. (2.1-2), (2.1-3), and (2.1-4) is a special case of the general *linear regression* problem. In the general case, unknown variables $x_1, x_2, \ldots, x_n$ are related to observable variables $\hat{b}_1, \hat{b}_2, \ldots, \hat{b}_m$ by

$$\hat{\mathbf{b}} = \mathbf{A}\mathbf{x} + \hat{\boldsymbol{\epsilon}} \tag{2.1-5}$$

where $\mathbf{A}$ is an $m \times n$ matrix of known constant elements and $\hat{\boldsymbol{\epsilon}}$ is a stochastic m-vector of observational errors. The error vector $\hat{\boldsymbol{\epsilon}}$ satisfies

$$E(\hat{\boldsymbol{\epsilon}}) = \mathbf{0}, \qquad E(\hat{\boldsymbol{\epsilon}}\hat{\boldsymbol{\epsilon}}^T) = \mathbf{S}, \tag{2.1-6}$$

where $\mathbf{S}$ is the $m \times m$ symmetric, positive-definite variance matrix of the observed vector $\hat{\mathbf{b}}$; that is,

$$\mathbf{S} = \boldsymbol{\Sigma}_{\hat{\mathbf{b}}} = E[(\hat{\mathbf{b}} - E(\hat{\mathbf{b}}))(\hat{\mathbf{b}} - E(\hat{\mathbf{b}}))^T]. \tag{2.1-7}$$

If all the $\hat{b}_j$ are uncorrelated, then $\mathbf{S}$ reduces to a diagonal matrix, and if in addition they all have the same variance σ^2, then

$$\mathbf{S} = \sigma^2 \mathbf{I}.$$

An alternate way of writing Eqs. (2.1-5) and (2.1-6) is

$$E(\hat{\mathbf{b}}) = \mathbf{A}\mathbf{x}, \tag{2.1-8}$$

$$\boldsymbol{\Sigma}_{\hat{\mathbf{b}}} = E[(\hat{\mathbf{b}} - \mathbf{A}\mathbf{x})(\hat{\mathbf{b}} - \mathbf{A}\mathbf{x})^T] = \mathbf{S}. \tag{2.1-9}$$

The *classical linear regression model* assumes that $m \geqslant n$ and that the matrix $\mathbf{A}$ has linearly independent columns; that is, that rank $(\mathbf{A}) = n$. Both of these restrictions will be dropped later, but first we will consider the classical problem.

2.2 Point Estimation

In the classical linear regression model the vector $\mathbf{x}$ is a unique, fixed, "true" but unknown vector that could, in principle, be determined from a complete knowledge of the distributions of the vectors $\hat{\mathbf{b}}$ and $\hat{\boldsymbol{\epsilon}}$. In actual practice, we do not have complete knowledge of these distributions but only knowledge of particular observed values for each of the $\hat{b}_i$. Therefore, we cannot determine the "true" values of the x_i–we can only estimate them. But before trying to estimate the vector $\mathbf{x}$, we consider the somewhat simpler problem of estimating the scalar function

$$\phi = c_1 x_1 + c_2 x_2 + \ldots + c_n x_n = \mathbf{c}^T \mathbf{x} \tag{2.2-1}$$

where the c_j are known constants. For this purpose, we seek a scalar function of the form

$$\hat{\phi} = u_1 \hat{b}_1 + u_2 \hat{b}_2 + \ldots + u_m \hat{b}_m = \mathbf{u}^T \hat{\mathbf{b}} \tag{2.2-2}$$

to use as an *estimator* of ϕ. This function is linear in the observations $\hat{b}_i$ and hence is called a *linear estimator.* The problem is to pick the coefficients u_i so that $\hat{\phi}$ gives a good estimate of ϕ.

The estimator $\hat{\phi}$ is said to be *unbiased* if its expectation is identically equal to ϕ; that is, if

$$E(\mathbf{u}^T \hat{\mathbf{b}}) \equiv \mathbf{c}^T \mathbf{x} \tag{2.2-3}$$

independently of the value of the unknown vector $\mathbf{x}$. Now

$$E(\mathbf{u}^T \hat{\mathbf{b}}) = \mathbf{u}^T E(\hat{\mathbf{b}}) = \mathbf{u}^T \mathbf{A}\mathbf{x},$$

so for an unbiased estimator,

$$\mathbf{u}^T \mathbf{A}\mathbf{x} \equiv \mathbf{c}^T \mathbf{x}$$

independently of $\mathbf{x}$, and this requires that

$$\mathbf{u}^T \mathbf{A} = \mathbf{c}^T. \tag{2.2-4}$$

Unbiasedness is certainly a desirable property for a good estimator $\hat{\phi}$, but if $m > n$, there will be many vectors $\mathbf{u}$ that satisfy (2.2-4) and hence many unbiased estimators for ϕ.

Another desirable property for a good estimator $\hat{\phi}$ is that it have a small dispersion or variance. Therefore we seek to choose, from the set of vectors $\mathbf{u}$ that satisfy (2.2-4), that particular $\mathbf{u}$ which gives the minimum variance for the estimator $\hat{\phi}$. The resulting estimator is called the *minimum-variance unbiased linear estimator* or simply the *best linear unbiased estimator.* The variance of the estimator $\hat{\phi}$ is

$$\sigma^2(\hat{\phi}) = E[(\hat{\phi} - E(\hat{\phi}))^2] = E[(\mathbf{u}^T\hat{\mathbf{b}} - E(\mathbf{u}^T\hat{\mathbf{b}}))^2] = E[(\mathbf{u}^T\hat{\mathbf{b}} - \mathbf{u}^T\mathbf{A}\mathbf{x})^2]$$
$$= \mathbf{u}^T E[(\hat{\mathbf{b}} - \mathbf{A}\mathbf{x})(\hat{\mathbf{b}} - \mathbf{A}\mathbf{x})^T]\mathbf{u}$$

or

$$\sigma^2(\hat{\phi}) = \mathbf{u}^T\mathbf{S}\mathbf{u}. \tag{2.2-5}$$

Therefore, to get the best linear unbiased estimator, we must choose that vector $\mathbf{u}$ which minimizes the quantity $\mathbf{u}^T\mathbf{S}\mathbf{u}$ subject to the constraint

$$\mathbf{u}^T\mathbf{A} - \mathbf{c}^T = \mathbf{0}.$$

Using the method of Lagrange, we seek to minimize the expression

$$L = \mathbf{u}^T\mathbf{S}\mathbf{u} + (\mathbf{u}^T\mathbf{A} - \mathbf{c}^T)\mathbf{p}, \tag{2.2-6}$$

where $\mathbf{p}$ is the n-vector of Lagrange multipliers. Differentiating with respect to the vectors $\mathbf{u}$ and $\mathbf{p}$ and equating the derivatives to zero gives

$$\frac{\partial L}{\partial \mathbf{u}} = 2\mathbf{S}\mathbf{u} + \mathbf{A}\mathbf{p} = \mathbf{0}, \tag{2.2-7}$$

$$\frac{\partial L}{\partial \mathbf{p}} = \mathbf{A}^T\mathbf{u} - \mathbf{c} = \mathbf{0}. \tag{2.2-8}$$

Solving (2.2-7) for $\mathbf{u}$ gives

$$\mathbf{u} = -\tfrac{1}{2}\mathbf{S}^{-1}\mathbf{A}\mathbf{p}, \tag{2.2-9}$$

and multiplying this equation on the left by $\mathbf{A}^T$ gives

$$\mathbf{A}^T\mathbf{u} = -\tfrac{1}{2}\mathbf{A}^T\mathbf{S}^{-1}\mathbf{A}\mathbf{p},$$

which by (2.2-8) gives

$$\mathbf{p} = -2(\mathbf{A}^T\mathbf{S}^{-1}\mathbf{A})^{-1}\mathbf{c}.$$

(The nonsingularity of $\mathbf{A}^T\mathbf{S}^{-1}\mathbf{A}$ follows from the linear independence of the columns of $\mathbf{A}$.) If this last result is combined with Eq. (2.2-9), the resulting solution is

$$\mathbf{u} = \mathbf{S}^{-1}\mathbf{A}(\mathbf{A}^T\mathbf{S}^{-1}\mathbf{A})^{-1}\mathbf{c}, \tag{2.2-10}$$

and the best linear unbiased estimate is

$$\hat{\phi} = \mathbf{c}^T(\mathbf{A}^T\mathbf{S}^{-1}\mathbf{A})^{-1}\mathbf{A}^T\mathbf{S}^{-1}\hat{\mathbf{b}}. \tag{2.2-11}$$

We now consider the problem of finding a *vector estimator* for the vector $\mathbf{x}$ itself. We seek a linear vector function of the form

$$\hat{\mathbf{x}} = \mathbf{U}^T \hat{\mathbf{b}} \tag{2.2-12}$$

where $\mathbf{U}^T$ is an $n \times m$ matrix to be determined. Such an estimator is said to be unbiased if

$$E(\hat{\mathbf{x}}) = \mathbf{x} \tag{2.2-13}$$

whatever the value of $\mathbf{x}$; that is, if

$$\mathbf{U}^T \mathbf{A}\mathbf{x} = \mathbf{x}$$

independently of $\mathbf{x}$. Thus, to assure that $\hat{\mathbf{x}}$ be unbiased, we must choose a matrix $\mathbf{U}$ that satisfies

$$\mathbf{U}^T \mathbf{A} = \mathbf{I}. \tag{2.2-14}$$

The variance matrix of the linear unbiased estimator $\hat{\mathbf{x}}$ is

$$\begin{aligned}\mathbf{\Sigma}_{\hat{\mathbf{x}}} &= E[(\hat{\mathbf{x}} - \mathbf{x})(\hat{\mathbf{x}} - \mathbf{x})^T] = E[(\mathbf{U}^T \hat{\mathbf{b}} - \mathbf{U}^T \mathbf{A}\mathbf{x})(\mathbf{U}^T \hat{\mathbf{b}} - \mathbf{U}^T \mathbf{A}\mathbf{x})^T] \\ &= \mathbf{U}^T E[(\hat{\mathbf{b}} - \mathbf{A}\mathbf{x})(\hat{\mathbf{b}} - \mathbf{A}\mathbf{x})^T]\mathbf{U}\end{aligned}$$

or

$$\mathbf{\Sigma}_{\hat{\mathbf{x}}} = \mathbf{U}^T \mathbf{S}\mathbf{U}. \tag{2.2-15}$$

The estimator $\hat{\mathbf{x}}$ is said to be the minimum-variance linear unbiased estimator, or the best linear unbiased estimator, if for any other linear unbiased estimator

$$\check{\mathbf{x}} = \mathbf{W}^T \hat{\mathbf{b}}$$

with variance matrix

$$\mathbf{\Sigma}_{\check{\mathbf{x}}} = \mathbf{W}^T \mathbf{S}\mathbf{W},$$

the difference

$$\mathbf{\Sigma}_{\check{\mathbf{x}}} - \mathbf{\Sigma}_{\hat{\mathbf{x}}} = \mathbf{W}^T \mathbf{S}\mathbf{W} - \mathbf{U}^T \mathbf{S}\mathbf{U}$$

is a nonnegative definite matrix. The determinant of the variance matrix of a vector estimator is called the *generalized variance* of the estimator, and the best linear unbiased estimator has the smallest generalized variance of all linear unbiased estimators. Also, if $\hat{\mathbf{x}}$ is the best linear unbiased estimator of $\mathbf{x}$, then each of the elements $\hat{x}_i$ is separately the minimum-variance linear unbiased estimator for the corresponding x_i. Therefore, to determine the best linear unbiased estimator for $\mathbf{x}$, we seek the best linear unbiased estimators $\hat{x}_i$ for each of the x_i separately.

Let the ith column of the identity matrix be denoted by $\mathbf{e}_i$, that is,

$$\mathbf{I}_n = (\mathbf{e}_1, \mathbf{e}_2, \ldots, \mathbf{e}_n),$$

and consider the linear functions

$$\phi_i = \mathbf{e}_i^T \mathbf{x} = x_i, \qquad i = 1, 2, \ldots, n.$$

The problem is to find the best linear unbiased estimators of the form

$$\hat{\phi}_i = \hat{x}_i = \mathbf{u}_i^T \hat{\mathbf{b}}, \qquad i = 1, 2, \ldots, n;$$

the solution is, by Eq. (2.2-11),

$$\hat{x}_i = \mathbf{e}_i^T (\mathbf{A}^T \mathbf{S}^{-1} \mathbf{A})^{-1} \mathbf{A}^T \mathbf{S}^{-1} \hat{\mathbf{b}}, \qquad i = 1, 2, \ldots, n.$$

These separate estimators are the elements of the vector estimator

$$\hat{\mathbf{x}} = (\mathbf{A}^T \mathbf{S}^{-1} \mathbf{A})^{-1} \mathbf{A}^T \mathbf{S}^{-1} \hat{\mathbf{b}}, \tag{2.2-16}$$

which is the best linear unbiased estimator of $\mathbf{x}$. The matrix $\mathbf{U}^T$ is

$$\mathbf{U}^T = (\mathbf{A}^T \mathbf{S}^{-1} \mathbf{A})^{-1} \mathbf{A}^T \mathbf{S}^{-1}, \tag{2.2-17}$$

and, by Eq. (2.2-15), the variance matrix of $\hat{\mathbf{x}}$ is

$$\mathbf{\Sigma}_{\hat{\mathbf{x}}} = (\mathbf{A}^T \mathbf{S}^{-1} \mathbf{A})^{-1}. \tag{2.2-18}$$

The estimator given in Eq. (2.2-11) for the scalar function

$$\phi = \mathbf{c}^T \mathbf{x}$$

can now be expressed as

$$\hat{\phi} = \mathbf{c}^T \hat{\mathbf{x}},$$

where $\hat{\mathbf{x}}$ is the vector estimator that we have just derived.

The estimators given by Eqs. (2.2-11) and (2.2-16) are sometimes called *Markov estimators* because Markov was once credited with giving for them the first justification not using the unnecessary assumption that the $\hat{b}_i$ be normally distributed. But the same estimators can be obtained by the *method of least squares,* which was originally developed by Gauss, who is now also credited with giving a development that did not use the assumption of normality [1]. This procedure seeks to pick the vector $\hat{\mathbf{x}}$ that minimizes the quadratic form

$$\theta(\hat{\mathbf{x}}) = (\hat{\mathbf{b}} - \mathbf{A}\hat{\mathbf{x}})^T \mathbf{S}^{-1} (\hat{\mathbf{b}} - \mathbf{A}\hat{\mathbf{x}}). \tag{2.2-19}$$

For a minimum, we require

$$\frac{\partial \theta}{\partial \hat{\mathbf{x}}} = -2\mathbf{A}^T \mathbf{S}^{-1} \hat{\mathbf{b}} + 2\mathbf{A}^T \mathbf{S}^{-1} \mathbf{A}\hat{\mathbf{x}} = \mathbf{0}$$

or

$$\mathbf{A}^T \mathbf{S}^{-1} \mathbf{A}\hat{\mathbf{x}} = \mathbf{A}^T \mathbf{S}^{-1} \hat{\mathbf{b}}. \tag{2.2-20}$$

The equations in (2.2-20) are called the *normal equations*; their solution is given by

$$\hat{\mathbf{x}} = (\mathbf{A}^T \mathbf{S}^{-1} \mathbf{A})^{-1} \mathbf{A}^T \mathbf{S}^{-1} \hat{\mathbf{b}},$$

which is the same as (2.2-16).

The computation of Markov estimates requires an almost complete knowledge of the variance matrix **S**. It is not necessary to know **S** exactly, for if the expression

$$\mathbf{M} = \alpha \mathbf{S}$$

denotes any scalar multiple of **S**, then

$$(\mathbf{A}^T \mathbf{M}^{-1} \mathbf{A})^{-1} \mathbf{A}^T \mathbf{M}^{-1} = (\mathbf{A}^T \mathbf{S}^{-1} \mathbf{A})^{-1} \mathbf{A}^T \mathbf{S}^{-1}.$$

But in many cases, **S** is not known to within a scalar multiple, and so the Markov estimate cannot be computed. The procedure often used in these cases is that of computing a *weighted least squares estimate*

$$\hat{\mathbf{x}} = (\mathbf{A}^T \mathbf{W}^{-1} \mathbf{A})^{-1} \mathbf{A}^T \mathbf{W}^{-1} \hat{\mathbf{b}}, \tag{2.2-21}$$

which minimizes the quadratic form

$$\theta(\hat{\mathbf{x}}) = (\hat{\mathbf{b}} - \mathbf{A}\hat{\mathbf{x}})^T \mathbf{W}^{-1} (\hat{\mathbf{b}} - \mathbf{A}\hat{\mathbf{x}}) \tag{2.2-22}$$

where **W** is a positive-definite weighting matrix used as a guess for **S**. The expression in (2.2-21) is an unbiased estimator that may or may not be close to the Markov minimum-variance estimator, depending on the matrix **W**. The problem of comparing the variances of weighted least squares estimators and of Markov estimators has been treated in papers by Magness and McGuire [2] and by Golub [3].

2.3 Interval Estimation

The estimates given by Eqs. (2.2-11), (2.2-16), and (2.2-21) are all called *point estimates* because, for a particular value of $\hat{\mathbf{b}}$, they give only single estimates $\hat{\phi}$ or $\hat{\mathbf{x}}$ without any information about the probable error of the estimate. In this section we will consider *interval estimates* that give probable upper and lower bounds for the quantities being estimated. As a simple example of interval estimation, consider the problem of estimating the mean value μ of a random variable $\hat{\xi}$ that has a normal distribution with known variance σ^2, that is,

$$f(\xi) = \frac{1}{\sigma} (2\pi)^{-1/2} \exp\left\{-\left[\frac{(\xi - \mu)^2}{2\sigma^2}\right]\right\}, \qquad -\infty < \xi < \infty. \tag{2.3-1}$$

Let $\hat{\xi}_1, \hat{\xi}_2, \ldots, \hat{\xi}_n$ be the sample random variables representing n successive samples from the ξ-distribution and consider the statistic

$$\bar{\xi} = \frac{1}{n} \sum_{i=1}^{n} \hat{\xi}_i, \tag{2.3-2}$$

which is called the *sample mean.* Each of the random variables $\hat{\xi}_i$ has a normal distribution with mean μ and variance σ^2, and we assume that the samples are independent of one another. It follows then [4, pp. 138–139] that the statistic $\bar{\xi}$ is normally distributed with mean μ and variance σ^2/n, and thus the random variable

$$\hat{\eta} = \frac{\bar{\xi} - \mu}{\sigma/\sqrt{n}}$$

has the standard normal distribution with mean 0 and variance 1. Therefore for any positive constant κ, the probability (before the samples are actually taken) that $\hat{\eta}$ will be included in the interval $[-\kappa, \kappa]$ is

$$\Pr\left\{-\kappa \leqslant \frac{\bar{\xi} - \mu}{\sigma/n} \leqslant \kappa\right\} = (2\pi)^{-1/2} \int_{-\kappa}^{\kappa} \exp\left[-\frac{\eta^2}{2}\right] d\eta.$$

This last equation can be written

$$\Pr\left\{\bar{\xi} - \kappa\frac{\sigma}{n} \leqslant \mu \leqslant \bar{\xi} + \kappa\frac{\sigma}{n}\right\} = \alpha \tag{2.3-3}$$

where

$$\alpha = (2\pi)^{-1/2} \int_{-\kappa}^{\kappa} \exp\left[-\frac{\eta^2}{2}\right] d\eta. \tag{2.3-4}$$

The interval

$$I_\kappa \equiv \left[\bar{\xi} - \kappa\frac{\sigma}{n}, \bar{\xi} + \kappa\frac{\sigma}{n}\right] \tag{2.3-5}$$

is a *confidence interval* for the mean value μ and the probability α is a measure of the confidence level associated with that interval. Before the samples are taken, α is, for a fixed value of κ, the probability that the interval I_κ will contain the true value of the mean μ. Of course, when the samples are actually taken and the sample mean $\bar{\xi}$ is computed, the interval I_κ becomes a fixed definite interval that either does or does not contain the unknown value μ. But if this procedure were carried out a very large number of times, then α would be the fractional number of times that the resulting interval would contain μ, so if the value of α is very close to 1, we can feel fairly confident that the particular interval I_κ obtained from the samples actually taken does contain the unknown mean. For a given fixed value of α, the corresponding value of κ, together with the values of σ^2 and n, determines the width of the corresponding confidence interval. As the value of α gets closer and closer to 1, the corresponding value of κ becomes larger and larger because larger and larger intervals are required to assure the α-level of confidence. A few values for κ together with the corresponding values for α are given in Table 2.1.

TABLE 2.1

κ	α
1.0	0.6827
2.0	0.9545
3.0	0.9973
4.0	0.9999

Now we consider again the problem of estimating the linear function

$$\phi = \mathbf{c}^T \mathbf{x} \tag{2.3-6}$$

where $\mathbf{x}$ is the unknown vector in the general linear regression model

$$\hat{\mathbf{b}} = \mathbf{A}\mathbf{x} + \hat{\boldsymbol{\epsilon}}, \qquad E(\hat{\boldsymbol{\epsilon}}) = \mathbf{0}, \qquad E(\hat{\boldsymbol{\epsilon}}\hat{\boldsymbol{\epsilon}}^T) = \boldsymbol{\Sigma}_{\hat{\mathbf{b}}} = \mathbf{S}, \tag{2.3-7}$$

with $\mathbf{S}$ a positive-definite matrix. In the last section, we obtained a point estimate of the form

$$\hat{\phi} = \mathbf{u}^T \hat{\mathbf{b}}, \tag{2.3-8}$$

which could also be expressed in the form

$$\hat{\phi} = \mathbf{c}^T \hat{\mathbf{x}} \tag{2.3-9}$$

where, for a given $\hat{\mathbf{b}}$, $\hat{\mathbf{x}}$ is the particular value of the vector $\mathbf{z}$ that minimizes the quadratic form

$$\theta(\mathbf{z}) = (\hat{\mathbf{b}} - \mathbf{A}\mathbf{z})^T \mathbf{S}^{-1} (\hat{\mathbf{b}} - \mathbf{A}\mathbf{z}). \tag{2.3-10}$$

The parameter ϕ that we are trying to estimate is the mean value of the random variable $\hat{\phi}$ that has variance

$$\sigma^2 (\hat{\phi}) = \mathbf{u}^T \mathbf{S}\mathbf{u}. \tag{2.3-11}$$

The reduced random variable

$$\hat{\rho} = \frac{\hat{\phi} - \phi}{\sigma(\hat{\phi})} \tag{2.3-12}$$

has mean 0 and variance 1, and if we know enough about the joint probability distribution of the random variables $\hat{b}_1, \hat{b}_2, \ldots, \hat{b}_m$ to determine the probability distribution function of $\hat{\rho}$—call it $f(\rho)$—then for any positive constant κ it follows that

$$\Pr\left\{-\kappa \leqslant \frac{\hat{\phi} - \phi}{\sigma(\hat{\phi})} \leqslant \kappa\right\} = \int_{-\kappa}^{\kappa} f(\rho)\, d\rho$$

or

$$\Pr\{\hat{\phi} - \kappa\sigma(\hat{\phi}) \leqslant \phi \leqslant \hat{\phi} + \kappa\sigma(\hat{\phi})\} = \alpha \qquad (2.3\text{-}13)$$

where

$$\alpha = \int_{-\kappa}^{\kappa} f(\rho)\, d\rho. \qquad (2.3\text{-}14)$$

In the general case, it is an extremely difficult or impossible task to determine an explicit formula for $f(\rho)$. *We therefore make the additional assumption that the random variables $b_1, b_2, \ldots, b_m$ are jointly distributed with a multivariate normal distribution* and write

$$\hat{\mathbf{b}} \sim N(\mathbf{Ax}, \mathbf{S}) \qquad (2.3\text{-}15)$$

to denote that the vector $\hat{\mathbf{b}}$ is normally distributed with expectation $\mathbf{Ax}$ and variance matrix $\mathbf{S}$. The distribution function for $\hat{\mathbf{b}}$ can be written

$$g(\hat{\mathbf{b}}) = [(2\pi)^m |\mathbf{S}|]^{-1/2} \exp\left[-\tfrac{1}{2}(\hat{\mathbf{b}} - \mathbf{Ax})^T \mathbf{S}^{-1} (\hat{\mathbf{b}} - \mathbf{Ax})\right] \qquad (2.3\text{-}16)$$

where $|\mathbf{S}|$ denotes the determinant of $\mathbf{S}$. The model obtained by adding this normality assumption to the assumptions in (2.3-7) is sometimes called the *classical multivariate normal linear regression model.* In this model the random variable $\hat{\rho}$ will be normally distributed. To see this we first note that $\mathbf{S}$, being positive definite, can be factored in the form

$$\mathbf{S} = \mathbf{Q\Lambda Q}^T$$

where $\mathbf{Q}$ is an $m \times m$ orthogonal matrix whose columns are eigenvectors of $\mathbf{S}$, and $\mathbf{\Lambda}$ is a diagonal matrix whose diagonal elements are the eigenvalues of $\mathbf{S}$ (which are all positive). If $\hat{\mathbf{d}}$ is a new random variable defined by

$$\hat{\mathbf{d}} = \mathbf{Q}^T \hat{\mathbf{b}},$$

then it can be shown [5, p. 19] that $\hat{\mathbf{d}}$ is normally distributed with mean $\mathbf{Q}^T\mathbf{Ax}$ and variance matrix $\mathbf{Q}^T\mathbf{SQ} = \mathbf{\Lambda}$; that is,

$$\hat{\mathbf{d}} \sim N(\mathbf{Q}^T \mathbf{Ax}, \mathbf{\Lambda}).$$

The random variable $\hat{\phi}$ expressed in terms of $\hat{\mathbf{d}}$ is

$$\hat{\phi} = \mathbf{u}^T \mathbf{Q}\hat{\mathbf{d}} = \sum_{i=1}^{m} (\mathbf{u}^T \mathbf{Q})_i \hat{d}_i$$

where $(\mathbf{u}^T\mathbf{Q})_i$ denotes the ith element of the row vector $\mathbf{u}^T\mathbf{Q}$. Since the variance matrix $\mathbf{\Lambda}$ is diagonal, the variables $\hat{d}_1, \hat{d}_2, \ldots, \hat{d}_m$ are stochastically independent and $\hat{\phi}$, being a linear combination of independently normally distributed random variables, is itself normally distributed. Therefore $\hat{\rho}$ has the standard normal distribution, with mean 0 and variance 1, and the function $f(\rho)$ is

$$f(\rho) = (2\pi)^{-1/2} \exp[-(1/2)\rho^2]. \tag{2.3-17}$$

Thus the relationship between α and κ is the same as that in Eq. (2.3-4) and the values in Table 2.1 satisfy Eq. (2.3-13).

Although the assumption of normality makes it easy to compute the confidence level α, it is not an essential requirement for the estimation of confidence intervals. The crucial step in determining the interval in (2.3-13) was finding the random variable $\hat{\rho}$, which is a function of the sample random variables and the unknown parameter ϕ, but whose distribution function $f(\rho)$ is independent of ϕ, and from which the end points of the interval can be expressed as functions of the sample variables, $\hat{\phi} \pm \kappa\sigma(\hat{\phi})$, independently of ϕ. Mood and Graybill [6, p. 256] give a general method for finding confidence intervals even when no such random variable $\hat{\rho}$ can be found. In many applications, however, the $\hat{b}_i$ actually are normally distributed and thus the assumption of normality is not only convenient, but is also *necessary*.

The confidence interval

$$I_\kappa(\hat{\phi}) \equiv [\hat{\phi} - \kappa\sigma(\hat{\phi}), \hat{\phi} + \kappa\sigma(\hat{\phi})] = [\mathbf{u}^T\hat{\mathbf{b}} - \kappa(\mathbf{u}^T\mathbf{S}\mathbf{u})^{1/2}, \mathbf{u}^T\hat{\mathbf{b}} + \kappa(\mathbf{u}^T\mathbf{S}\mathbf{u})^{1/2}] \tag{2.3-18}$$

has an interesting geometrical interpretation. For a particular fixed value of $\hat{\mathbf{b}}$, Eq. (2.3-10) is the equation of a quadratic function, which we call the *residual surface.* The minimum value of this function is

$$r_0 = (\hat{\mathbf{b}} - \mathbf{A}\hat{\mathbf{x}})^T \mathbf{S}^{-1} (\hat{\mathbf{b}} - \mathbf{A}\hat{\mathbf{x}}) \tag{2.3-19}$$

where the vector $\hat{\mathbf{x}}$ satisfies the normal equations,

$$\mathbf{A}^T \mathbf{S}^{-1} \mathbf{A}\hat{\mathbf{x}} = \mathbf{A}^T \mathbf{S}^{-1} \hat{\mathbf{b}},$$

from which it follows that

$$r_0 = \hat{\mathbf{b}}^T \mathbf{S}^{-1} \hat{\mathbf{b}} - \hat{\mathbf{x}}^T \mathbf{A}^T \mathbf{S}^{-1} \mathbf{A}\hat{\mathbf{x}}. \tag{2.3-20}$$

Now suppose that κ is any positive constant and consider the equation

$$(\hat{\mathbf{b}} - \mathbf{A}\mathbf{z})^T \mathbf{S}^{-1} (\hat{\mathbf{b}} - \mathbf{A}\mathbf{z}) = r_0 + \kappa^2. \tag{2.3-21}$$

If $v_1, v_2, \ldots, v_n$ is a new set of variables defined by

$$\mathbf{z} = \hat{\mathbf{x}} + \mathbf{v}, \tag{2.3-22}$$

then Eq. (2.3-21) can be reduced to

$$\hat{\mathbf{b}}^T \mathbf{S}^{-1} \hat{\mathbf{b}} - \hat{\mathbf{x}}^T \mathbf{A}^T \mathbf{S}^{-1} \mathbf{A}\hat{\mathbf{x}} + \mathbf{v}^T \mathbf{A}^T \mathbf{S}^{-1} \mathbf{A}\mathbf{v} = r_0 + \kappa^2,$$

and if Eq. (2.3-20) is substituted into this expression, the result is

$$\mathbf{v}^T \mathbf{A}^T \mathbf{S}^{-1} \mathbf{A}\mathbf{v} = \kappa^2, \tag{2.3-23}$$

which is the equation of an ellipsoid centered at the origin of the v-space. The change of variables in (2.3-22) is simply a translation of the origin of the z-system to the point $\hat{\mathbf{x}}$. Applying the inverse translation to Eq. (2.3-23) gives

$$(\mathbf{z} - \hat{\mathbf{x}})^T \mathbf{A}^T \mathbf{S}^{-1} \mathbf{A}(\mathbf{z} - \hat{\mathbf{x}}) = \kappa^2, \tag{2.3-24}$$

which is the standard form of the equation of an ellipsoid centered at the point $\hat{\mathbf{x}}$ in the z-space. We will call this ellipsoid the *κ-ellipsoid.* The expression in (2.3-21) is just an alternate form of its equation. Figure 2.1 illustrates the geometrical relationship between the residual surface and a κ-ellipsoid when $n = 2$. The κ-ellipsoid E is the projection onto the z-plane of the $(r_0 + \kappa^2)$-contour on the residual surface $\theta(\mathbf{z})$.

The geometrical interpretation of the confidence interval $I_\kappa(\hat{\phi})$ involves the *planes of support* of the κ-ellipsoid. An excellent treatment of the geometry of ellipsoids and their planes of support has been given by Scheffe [7, pp. 406–411], who defined a plane of support as a plane such that the ellipsoid lies entirely on one side of and has at least one point in common with it. For a nondegenerate ellipsoid a plane of support is any plane that has exactly one point in common with the ellipsoid, and for any vector $\mathbf{h} \neq 0$ the ellipsoid will have two planes of support that are orthogonal to $\mathbf{h}$. If the equation of the ellipsoid is

$$(\mathbf{z} - \mathbf{a})^T \mathbf{M}(\mathbf{z} - \mathbf{a}) = \gamma^2 \tag{2.3-25}$$

where $\mathbf{a}$ is a constant vector, $\mathbf{M}$ is a symmetric positive-definite matrix, and γ^2 is any positive constant, then the equations of the two planes of support orthogonal to the vector $\mathbf{h}$ are

$$\mathbf{h}^T \mathbf{z} = \mathbf{h}^T \mathbf{a} \pm \gamma(\mathbf{h}^T \mathbf{M}^{-1} \mathbf{h})^{1/2}. \tag{2.3-26}$$

Now consider the κ-ellipsoid, Eq. (2.3-24), and the vector $\mathbf{c}$. We have already seen that

$$\hat{\phi} = \mathbf{u}^T \hat{\mathbf{b}} = \mathbf{c}^T \hat{\mathbf{x}}$$

where $\hat{\mathbf{x}}$ is the center of the κ-ellipsoid. The equations of the two planes of support orthogonal to $\mathbf{c}$ are

$$\mathbf{c}^T \mathbf{z} = \mathbf{c}^T \hat{\mathbf{x}} \pm \kappa[\mathbf{c}^T(\mathbf{A}^T \mathbf{S}^{-1} \mathbf{A})^{-1} \mathbf{c}]^{1/2}. \tag{2.3-27}$$

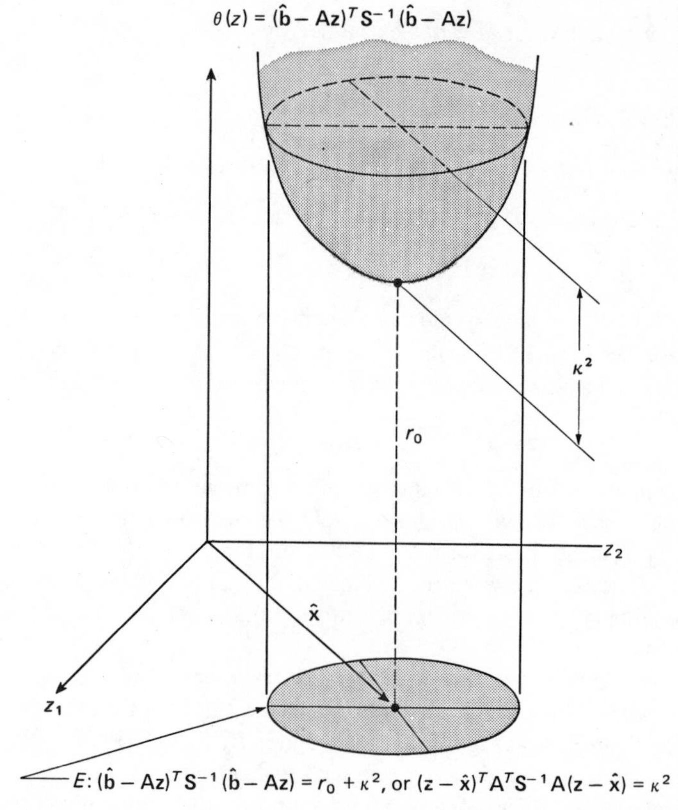

Figure 2.1. The residual surface and a κ-ellipsoid for a problem with $n = 2$.

It follows from Eq. (2.2-10) that the variance of $\hat{\phi}$ can be expressed

$$\sigma^2(\hat{\phi}) = \mathbf{u}^T \mathbf{S}\mathbf{u} = \mathbf{c}^T(\mathbf{A}^T \mathbf{S}^{-1} \mathbf{A})^{-1} \mathbf{c}. \tag{2.3-28}$$

Therefore the confidence interval $I_\kappa(\hat{\phi})$ can be written

$$I_\kappa(\hat{\phi}) = \{\mathbf{c}^T \hat{\mathbf{x}} - \kappa[\mathbf{c}^T(\mathbf{A}^T \mathbf{S}^{-1} \mathbf{A})^{-1}\mathbf{c}]^{1/2}, \mathbf{c}^T \hat{\mathbf{x}} + \kappa[\mathbf{c}^T(\mathbf{A}^T \mathbf{S}^{-1} \mathbf{A})^{-1} \mathbf{c}]^{1/2}\} \tag{2.3-29}$$

and each of its end points is associated with one of the two support planes, the value of the end point being equal to the value assumed by the linear function $\mathbf{c}^T\mathbf{z}$ on the corresponding support plane. Figure 2.2 illustrates this relationship for a problem where $n = 2$. The support plane S_+ is the locus of all points $\mathbf{z}$ such that

$$\mathbf{c}^T \mathbf{z} = \mathbf{c}^T \hat{\mathbf{x}} + \kappa[\mathbf{c}^T(\mathbf{A}^T \mathbf{S}^{-1} \mathbf{A})^{-1} \mathbf{c}]^{1/2}$$

and this is the value of the upper end point of the confidence interval $I_\kappa(\hat{\phi})$. A similar statement can be made about the support plane S_- and the lower end point of $I_\kappa(\hat{\phi})$.

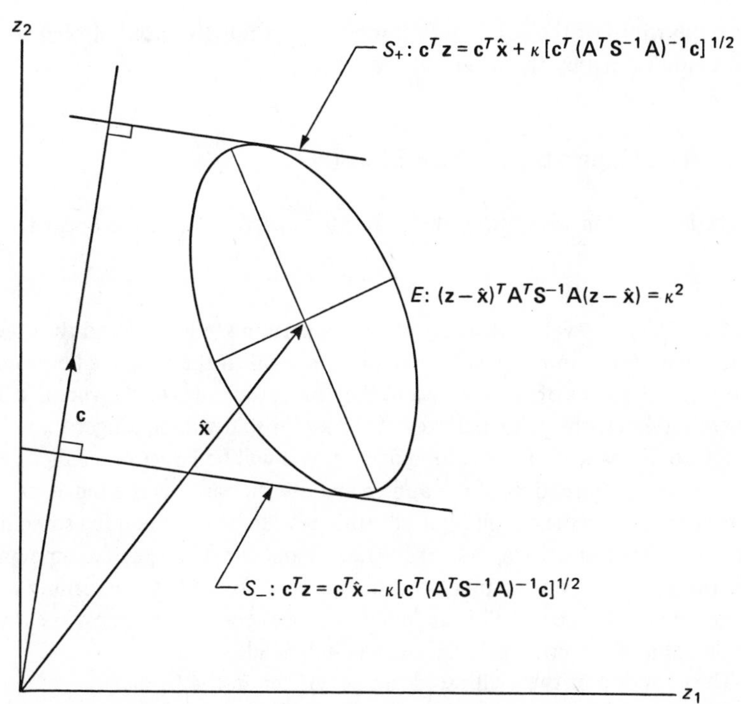

Figure 2.2. The κ-ellipsoid and support planes orthogonal to the vector **c** for a problem with $n = 2$.

The width of the confidence interval (2.3-29) is

$$2\kappa[\mathbf{c}^T(\mathbf{A}^T \mathbf{S}^{-1} \mathbf{A})^{-1} \mathbf{c}]^{1/2}$$

and this width is directly proportional to the distance between the two support planes defined by (2.3-27). If $\mathbf{h}_1$ and $\mathbf{h}_2$ are vectors that are collinear with **c** and that terminate on the planes S_- and S_+, respectively, then the distance between the two planes is

$$\|\mathbf{h}_2 - \mathbf{h}_1\|_2 = [(\mathbf{h}_2 - \mathbf{h}_1)^T (\mathbf{h}_2 - \mathbf{h}_1)]^{1/2},$$

which is just the Euclidean length of the difference vectors $(\mathbf{h}_2 - \mathbf{h}_1)$. From Eq. (2.3-27), it follows that

$$\mathbf{c}^T(\mathbf{h}_2 - \mathbf{h}_1) = 2\kappa[\mathbf{c}^T(\mathbf{A}^T \mathbf{S}^{-1} \mathbf{A})\mathbf{c}]^{1/2},$$

and since $\mathbf{c}$ is collinear with $(\mathbf{h}_2 - \mathbf{h}_1)$, the dot product on the left is the product of the lengths taken separately, so

$$\|\mathbf{c}\|_2 \, \|\mathbf{h}_2 - \mathbf{h}_1\|_2 = 2\kappa[\mathbf{c}^T(\mathbf{A}^T \mathbf{S}^{-1} \mathbf{A})\mathbf{c}]^{1/2}.$$

Thus the width of the confidence interval is just the distance between the support planes multiplied by the length of the vector $\mathbf{c}$.

2.4 The Geometry of the κ-Ellipsoid

In the last section we showed that, for interval estimates of the linear function

$$\phi = \mathbf{c}^T \mathbf{x}, \tag{2.4-1}$$

each confidence level α is associated with a certain κ-ellipsoid and that the width of the confidence interval is directly proportional to the distance between the two support planes of the κ-ellipsoid that are orthogonal to the vector $\mathbf{c}$. In general, some of the principal axes of the κ-ellipsoid will be longer than others; therefore, the width of the confidence interval will be larger or smaller depending on whether the direction of $\mathbf{c}$ is approximately the same as the longer or the shorter axes. This means that better estimates can be obtained for some linear functions than for others, or that the model and the data specify some linear functions better (or more precisely) than others. In order to determine which are the more closely specified linear functions, it is necessary to know the directions and lengths of the principal axes of the κ-ellipsoid.

The equation of the κ-ellipsoid can be written in the form

$$(\mathbf{z} - \hat{\mathbf{x}})^T \mathbf{A}^T \mathbf{S}^{-1} \mathbf{A}(\mathbf{z} - \hat{\mathbf{x}}) = \kappa^2 \tag{2.4-2}$$

where $\hat{\mathbf{x}}$ is the (unique) solution of the normal equations; that is,

$$(\mathbf{A}^T \mathbf{S}^{-1} \mathbf{A})\hat{\mathbf{x}} = \mathbf{A}^T \mathbf{S}^{-1} \hat{\mathbf{b}}. \tag{2.4-3}$$

In order to determine the directions of the principal axes, we make the linear transformation of coordinates

$$\mathbf{v} = \mathbf{z} - \hat{\mathbf{x}}, \tag{2.4-4}$$

which reduces the equation of the ellipsoid to

$$\mathbf{v}^T \mathbf{A}^T \mathbf{S}^{-1} \mathbf{A}\mathbf{v} = \kappa^2. \tag{2.4-5}$$

This transformation, illustrated for a two-dimensional ellipsoid in Fig. 2.3a, b, translates the center of the ellipsoid to the origin of the coordinate system but does not alter its shape or orientation in space. Therefore the length and direction of each of the principal axes is unchanged by the transformation. Since the ellipsoid is now centered at the origin, each of the axes goes through the origin and intersects the surface of the ellipsoid at two opposite points, each of which is characterized by the fact that the radius vector to that point is orthogonal to

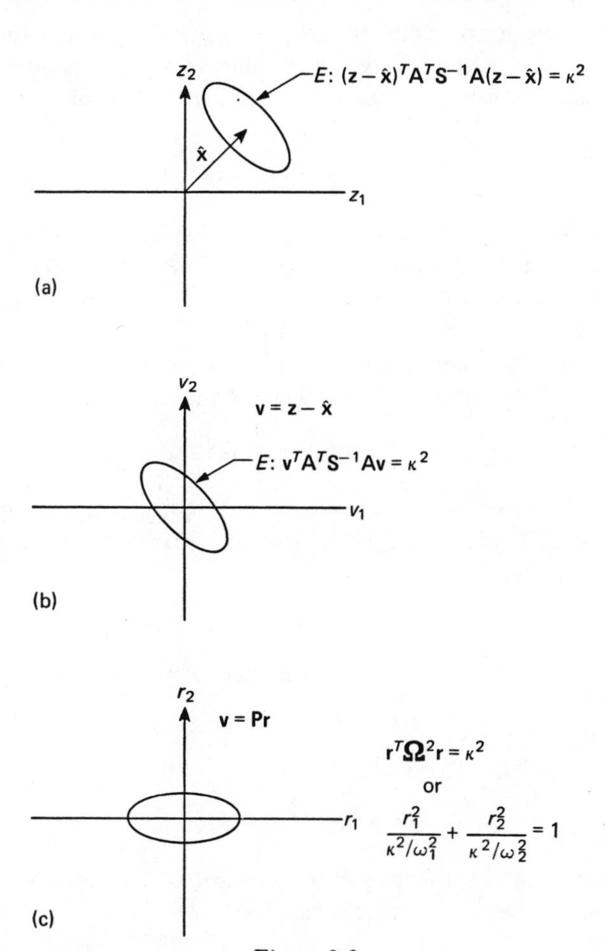

Figure 2.3.

the surface. More precisely, the radius vector to any point on the surface of the ellipsoid will be collinear with one of the principal axes if and only if it is orthogonal to the tangent plane through that point, or equivalently if it is in the same direction as the normal to the surface at that point. If the surface of the ellipsoid is considered to be an *isotimic* (equipotential) *surface* for the scalar field

$$\beta(\mathbf{v}) = \mathbf{v}^T \mathbf{A}^T \mathbf{S}^{-1} \mathbf{A}\mathbf{v},$$

then, for any point $\mathbf{v}$ on the surface, the direction of the gradient vector

$$\nabla\beta = 2\mathbf{A}^T\mathbf{S}^{-1}\mathbf{A}\mathbf{v}$$

is the same as the normal to the surface at that point. Therefore, the radius vector $\mathbf{p}$ to a point on the surface will be collinear with a principal axis if and only if the gradient vector at that point is a scalar multiple of $\mathbf{p}$; that is, if and only if

$$\mathbf{A}^T\mathbf{S}^{-1}\mathbf{A}\mathbf{p} = \lambda\mathbf{p} \tag{2.4-6}$$

for some constant λ. The solutions to this equation are the eigenvectors of the matrix $\mathbf{A}^T\mathbf{S}^{-1}\mathbf{A}$. Since this matrix is positive definite, it will have n *real* positive eigenvalues, which we denote by $\omega_1^2, \omega_2^2, \ldots, \omega_n^2$, with n corresponding mutually orthogonal eigenvectors $\mathbf{p}_1, \mathbf{p}_2, \ldots, \mathbf{p}_n$. Each of the n principal axes of the κ-ellipsoid has the same direction as one of these eigenvectors.

Suppose that the eigenvalues are ordered so that

$$\omega_1^2 \geqslant \omega_2^2 \geqslant \ldots \geqslant \omega_n^2 > 0$$

and that each of the corresponding eigenvectors $\mathbf{p}_1, \mathbf{p}_2, \ldots, \mathbf{p}_n$ is normalized to have unit length. Let $\mathbf{\Omega}^2$ be the $n \times n$ diagonal matrix with the eigenvalues as diagonal entries, that is,

$$\mathbf{\Omega}^2 = \operatorname{diag}(\omega_1^2, \omega_2^2, \ldots, \omega_n^2), \tag{2.4-7}$$

and let $\mathbf{P}$ be the matrix whose columns are the normalized eigenvectors, that is,

$$\mathbf{P} = (\mathbf{p}_1, \mathbf{p}_2, \ldots, \mathbf{p}_n). \tag{2.4-8}$$

Then the matrix $\mathbf{P}$ is orthogonal and

$$\mathbf{P}^T\mathbf{A}^T\mathbf{S}^{-1}\mathbf{A}\mathbf{P} = \mathbf{\Omega}^2. \tag{2.4-9}$$

In order to determine the lengths of the principal axes we make another change of variables. Let the new vector variable $\mathbf{r}$ be defined by

$$\mathbf{r} = \mathbf{P}^T\mathbf{v}. \tag{2.4-10}$$

Since $\mathbf{P}$ is an orthogonal matrix, this transformation corresponds to a rotation of the space about the origin, and it leaves the length of each vector in the space unchanged. Figure 2.3b, c illustrates this rotation for the two-dimensional case. The lengths of the principal axes of the ellipsoid defined by Eq. (2.4-5) are not changed by this transformation, but the effect of the rotation is to line them up with the axes of the coordinate system. To see this we solve Eq. (2.4-10) for $\mathbf{v}$,

$$\mathbf{v} = \mathbf{P}\mathbf{r},$$

and substitute this result into (2.4-5) to get

$$\mathbf{r}^T \mathbf{P}^T \mathbf{A}^T \mathbf{S}^{-1} \mathbf{A}\mathbf{P}\mathbf{r} = \kappa^2,$$

which by Eq. (2.4-9) reduces to

$$\mathbf{r}^T \mathbf{\Omega}^2 \mathbf{r} = \kappa^2. \tag{2.4-11}$$

This last equation can also be written

$$\omega_1^2 r_1^2 + \omega_2^2 r_2^2 + \ldots + \omega_n^2 r_n^2 = \kappa^2;$$

or

$$\frac{r_1^2}{\kappa^2/\omega_1^2} + \frac{r_2^2}{\kappa^2/\omega_2^2} + \ldots + \frac{r_n^2}{\kappa^2/\omega_n^2} = 1, \tag{2.4-12}$$

which is the standard form of the equation for an ellipsoid whose principal axes are collinear with the coordinate axes and have lengths

$$\Gamma_1 = \frac{2\kappa}{\omega_1}, \; \Gamma_2 = \frac{2\kappa}{\omega_2}, \ldots, \; \Gamma_n = \frac{2\kappa}{\omega_n}. \tag{2.4-13}$$

Thus, each of the principal axes of the κ-ellipsoid has the same direction as one of the eigenvectors of $\mathbf{A}^T\mathbf{S}^{-1}\mathbf{A}$, and the length of the axis is inversely proportional to the square root of the corresponding eigenvalue. The equations of the line segments that are the principal axes can be written

$$\mathbf{z}_j = \hat{\mathbf{x}} + \tau_j \frac{\kappa}{\omega_j} \mathbf{p}_j, \qquad -1 \leqslant \tau_j \leqslant 1, \qquad j = 1, 2, \ldots, n. \tag{2.4-14}$$

Since the width of the confidence interval for the linear function $\phi = \mathbf{c}^T\mathbf{x}$ is longer or shorter depending on whether $\mathbf{c}$ is in the same general direction as a longer or shorter principal axis, it follows that narrower intervals can be obtained for functions in which $\mathbf{c}$ is in the same general direction as an eigenvector corresponding to one of the larger eigenvalues than for functions that have $\mathbf{c}$ in the direction of an eigenvector corresponding to one of the smaller eigenvalues.

2.5 Poorly Conditioned Normal Equations

In the last section we saw that the size, shape, and orientation of the κ-ellipsoid are completely determined by the value of κ and the eigensystem of $\mathbf{A}^T\mathbf{S}^{-1}\mathbf{A}$. But the location of the κ-ellipsoid is dependent on the particular sample value of the vector $\hat{\mathbf{b}}$ that determines the vector $\hat{\mathbf{x}}$ through the normal equations (2.4-3), that is,

$$(\mathbf{A}^T \mathbf{S}^{-1} \mathbf{A})\hat{\mathbf{x}} = \mathbf{A}^T \mathbf{S}^{-1} \hat{\mathbf{b}}. \tag{2.5-1}$$

To see how the location of the center of the ellipsoid changes with the random error in $\hat{\mathbf{b}}$, we consider the error vector $\Delta\mathbf{x}$ defined by

$$\hat{\mathbf{x}} = \mathbf{x} + \Delta\mathbf{x} \tag{2.5-2}$$

where $\mathbf{x}$ is the true value of the unknown vector. Substituting this expression into Eq. (2.5-1) gives

$$(\mathbf{A}^T\mathbf{S}^{-1}\mathbf{A})(\mathbf{x} + \Delta\mathbf{x}) = \mathbf{A}^T\mathbf{S}^{-1}\hat{\mathbf{b}} = \mathbf{A}^T\mathbf{S}^{-1}(\mathbf{A}\mathbf{x} + \hat{\boldsymbol{\epsilon}}),$$

which reduces to

$$(\mathbf{A}^T\mathbf{S}^{-1}\mathbf{A})\Delta\mathbf{x} = \mathbf{A}^T\mathbf{S}^{-1}\hat{\boldsymbol{\epsilon}}. \tag{2.5-3}$$

A comparison of this equation with Eq. (2.5-1) shows that the relationship between the error in $\hat{\mathbf{x}}$ and the random sample error $\hat{\boldsymbol{\epsilon}}$ is the same as that between the estimated value $\hat{\mathbf{x}}$ and the measured value $\hat{\mathbf{b}}$.

Since $\mathbf{S}^{-1}$ is positive definite, there exists a symmetric nonsingular matrix $\mathbf{H}$ such that

$$\mathbf{S}^{-1} = \mathbf{H}^T\mathbf{H} = \mathbf{H}^2; \tag{2.5-4}$$

so Eq. (2.5-3) can be written

$$(\mathbf{A}^T\mathbf{H}^T\mathbf{H}\mathbf{A})\Delta\mathbf{x} = \mathbf{A}^T\mathbf{H}^T\mathbf{H}\hat{\boldsymbol{\epsilon}}. \tag{2.5-5}$$

We now introduce a factorization of the matrix $\mathbf{HA}$ that is analogous to the expansion for the kernel function $K(t,s)$ given in Eq. (1.2-16). Any $m \times n$ matrix $\mathbf{B}$, with $m > n$, can be factored in the form

$$\mathbf{B} = \mathbf{L}\boldsymbol{\Xi}\mathbf{R}^T \tag{2.5-6}$$

where $\mathbf{L}$ and $\mathbf{R}$ are square orthogonal matrices and $\boldsymbol{\Xi}$ is an $m \times \mathbf{n}$ matrix of the form

$$\boldsymbol{\Xi} = \begin{bmatrix} \mathbf{D} \\ \mathbf{0} \end{bmatrix}, \qquad \mathbf{D} = \operatorname{diag}(\xi_1, \xi_2, \ldots, \xi_n). \tag{2.5-7}$$

The lower submatrix of $\boldsymbol{\Xi}$ is the $(m-n) \times n$ zero matrix; the numbers $\xi_1, \xi_2, \ldots, \xi_n$ are the *singular values* of the matrix $\mathbf{B}$, which are the positive square roots of the eigenvalues of $\mathbf{B}^T\mathbf{B}$, that is,

$$\xi_i = +[\lambda_i(\mathbf{B}^T\mathbf{B})]^{1/2}, \qquad i = 1, 2, \ldots, n. \tag{2.5-8}$$

If the rank of the matrix $\mathbf{B}$ is $r < n$, then $(n-r)$ of these singular values will be equal to zero. The columns of the orthogonal matrix $\mathbf{L}$ are the eigenvectors of $\mathbf{B}\mathbf{B}^T$ and the columns of the matrix $\mathbf{R}$ (rows of $\mathbf{R}^T$) are the eigenvectors of $\mathbf{B}^T\mathbf{B}$. A similar factorization holds for the case where $m < n$, but in this case the matrix $\boldsymbol{\Xi}$ has the form

$$\boldsymbol{\Xi} = [\mathbf{D}, \mathbf{0}], \qquad \mathbf{D} = \operatorname{diag}(\xi_1, \xi_2, \ldots, \xi_m),$$

and the singular values are the positive square roots of the eigenvalues of $\mathbf{BB}^T$. Actually, any eigenvalue of $\mathbf{BB}^T$ will also be an eigenvalue of $\mathbf{B}^T\mathbf{B}$, but since $m < n$, there will be $(n-m)$ zero eigenvalues of $\mathbf{B}^T\mathbf{B}$ that are not eigenvalues of $\mathbf{BB}^T$. In the case where $m = n$,

$$\mathbf{\Xi} = \mathbf{D} = \text{diag}(\xi_1, \xi_2, \ldots, \xi_n)$$

and the eigenvalues of $\mathbf{B}^T\mathbf{B}$ are identical with those of $\mathbf{BB}^T$. Now the matrix $\mathbf{HA}$ is an $m \times n$ matrix with $m > n$ and rank $(\mathbf{A}) = n$, so its singular values are just the positive square roots of the eigenvalues of the matrix

$$(\mathbf{HA})^T \mathbf{HA} = \mathbf{A}^T \mathbf{H}^T \mathbf{HA} = \mathbf{A}^T \mathbf{S}^{-1} \mathbf{A}.$$

The eigenvalues of this matrix are $\omega_1^2, \omega_2^2, \ldots, \omega_n^2$ and its eigenvectors are $\mathbf{p}_1$, $\mathbf{p}_2, \ldots, \mathbf{p}_n$. Therefore, the singular values of $\mathbf{HA}$ are $\omega_1 \geqslant \omega_2 \geqslant \cdots \geqslant \omega_n$; for the factorization of $\mathbf{HA}$,

$$\mathbf{\Xi} = \left(\frac{\mathbf{\Omega}}{\mathbf{0}}\right) \qquad \text{and} \qquad \mathbf{R} = \mathbf{P}$$

where

$$\mathbf{\Omega} = \text{diag}(\omega_1, \omega_2, \ldots, \omega_n) \tag{2.5-9}$$

is the "square root" of the matrix defined by Eq. (2.4-7) and $\mathbf{P}$ is the matrix defined in (2.4-8). If $\mathbf{L}$ is the orthogonal matrix whose columns are the eigenvectors of $(\mathbf{HA})(\mathbf{HA})^T = \mathbf{HAA}^T\mathbf{H}$, then the factorization can be written

$$\mathbf{HA} = \mathbf{L}\left(\frac{\mathbf{\Omega}}{\mathbf{0}}\right)\mathbf{P}^T. \tag{2.5-10}$$

If this expression is substituted into Eq. (2.5-5), the result is

$$\mathbf{P\Omega}^2 \mathbf{P}^T \Delta\mathbf{x} = \mathbf{P}(\mathbf{\Omega}, \mathbf{0}) \mathbf{L}^T \mathbf{H}\hat{\boldsymbol{\epsilon}}$$

or

$$\mathbf{\Omega}^2 \mathbf{P}^T \Delta\mathbf{x} = (\mathbf{\Omega}, \mathbf{0}) \mathbf{L}^T \mathbf{H}\hat{\boldsymbol{\epsilon}}.$$

Since none of the eigenvalues $\omega_1^2, \omega_2^2, \ldots, \omega_n^2$ are equal to zero, the matrix $\mathbf{\Omega}^2$ is nonsingular, and we can multiply both sides of the foregoing equation by $\mathbf{P\Omega}^{-2} = \mathbf{P}(\mathbf{\Omega}^{-1})^2$ to get

$$\Delta\mathbf{x} = \mathbf{P}(\mathbf{\Omega}^{-1}, \mathbf{0}) \mathbf{L}^T \mathbf{H}\hat{\boldsymbol{\epsilon}}, \tag{2.5-11}$$

which can also be written

$$\Delta\mathbf{x} = \frac{(\mathbf{L}^T \mathbf{H}\hat{\boldsymbol{\epsilon}})_1}{\omega_1}\mathbf{p}_1 + \frac{(\mathbf{L}^T \mathbf{H}\hat{\boldsymbol{\epsilon}})_2}{\omega_2}\mathbf{p}_2 + \ldots + \frac{(\mathbf{L}^T \mathbf{H}\hat{\boldsymbol{\epsilon}})_n}{\omega_n}\mathbf{p}_n, \tag{2.5-12}$$

where $(\mathbf{L}^T\mathbf{H}\hat{\boldsymbol{\epsilon}})_i$ denotes the ith element of the m-vector $\mathbf{L}^T\mathbf{H}\hat{\boldsymbol{\epsilon}}$. Note that the error depends only upon the first n elements of this vector.

Equation (2.5-12) expresses the error in $\mathbf{x}$, induced by the observational error $\hat{\boldsymbol{\epsilon}}$, as a linear combination of the eigenvectors of $\mathbf{A}^T\mathbf{S}^{-1}\mathbf{A}$. The component of the error along each of these eigenvectors is inversely proportional to the value of the corresponding singular value of $\mathbf{HA}$. Therefore, if the elements of the vector $\mathbf{L}^T\mathbf{H}\hat{\boldsymbol{\epsilon}}$ do not vary widely in magnitude, the components of the error will be greater along the eigenvectors corresponding to the smaller singular values than along those corresponding to the larger eigenvalues. This means that the center of the κ-ellipsoid will be displaced from the true $\mathbf{x}$ by greater amounts in the directions corresponding to the smaller singular values than in the directions corresponding to the larger ones. More precisely, if the error vector $\hat{\boldsymbol{\epsilon}}$ is distributed in such a way that the distribution of the vector $\mathbf{H}\hat{\boldsymbol{\epsilon}}$ is fairly regular, then the vector $\mathbf{L}^T\mathbf{H}\hat{\boldsymbol{\epsilon}}$ will also have a fairly regular distribution because multiplication by the orthogonal matrix $\mathbf{L}^T$ corresponds to a simple rotation of $\mathbf{H}\hat{\boldsymbol{\epsilon}}$, which does not change its length; thus, if there is a significant variation in the sizes of the singular values, it is most probable that the displacement of the center $\hat{\mathbf{x}}$ of the κ-ellipsoid from the true value $\mathbf{x}$ will be greater in the directions of the eigenvectors corresponding to the smaller singular values and lesser in the directions of the eigenvectors corresponding to the larger singular values. This greater uncertainty in the directions of the smaller singular values is reflected in Eq. (2.4-13), which shows that the lengths of the principal axes of the κ-ellipsoid are longer for those axes that are collinear with eigenvectors corresponding to the smaller singular values, and shorter for those that are collinear with eigenvectors corresponding to the larger singular values.

We say that the system of normal equations (2.5-1) is *poorly conditioned* if small relative changes in the vector $\hat{\mathbf{b}}$ produce large relative changes in the vector $\hat{\mathbf{x}}$. Equation (2.5-12) shows how changes in $\hat{\mathbf{b}}$ produce changes in $\hat{\mathbf{x}}$, but it relates the two absolute errors rather than the relative errors. To establish a relationship between the relative errors we solve Eq. (2.5-5) for $\Delta\mathbf{x}$ and consider the norm of the result, that is,

$$\|\Delta\mathbf{x}\| = \|(\mathbf{A}^T\mathbf{H}^T\mathbf{HA})^{-1}\mathbf{A}^T\mathbf{H}^T\mathbf{H}\hat{\boldsymbol{\epsilon}}\|.$$

We do not yet specify a particular vector norm, but if we use a *consistent matrix norm*, it follows that

$$\|\Delta\mathbf{x}\| \leqslant \|(\mathbf{A}^T\mathbf{H}^T\mathbf{HA})^{-1}\mathbf{A}^T\mathbf{H}^T\|\,\|\mathbf{H}\hat{\boldsymbol{\epsilon}}\|. \tag{2.5-13}$$

Readers who are not familiar with the basic properties of vector and matrix norms can find an excellent treatment of them in Householder's *The Theory of Matrices in Numerical Analysis* [8]. Corresponding to the true, but unknown, value of $\mathbf{x}$, there is a true value of $\mathbf{b}$ that is defined by

$$\mathbf{b} = \mathbf{Ax} = E(\hat{\mathbf{b}}). \tag{2.5-14}$$

Premultiplying by the matrix $\mathbf{H}$ and taking the norm of the result gives

$$\|\mathbf{Hb}\| = \|\mathbf{HAx}\|,$$

from which it follows that

$$\|\mathbf{Hb}\| \leqslant \|\mathbf{HA}\|\,\|\mathbf{x}\|. \tag{2.5-15}$$

Multiplying Eqs. (2.5-13) and (2.5-15) together gives

$$\|\Delta\mathbf{x}\|\,\|\mathbf{Hb}\| \leqslant \|(\mathbf{A}^T\mathbf{H}^T\mathbf{HA})^{-1}\mathbf{A}^T\mathbf{H}^T\|\,\|\mathbf{HA}\|\,\|\mathbf{H}\hat{\boldsymbol{\epsilon}}\|\,\|\mathbf{x}\|,$$

which can also be written

$$\frac{\|\Delta\mathbf{x}\|}{\|\mathbf{x}\|} \leqslant \|(\mathbf{A}^T\mathbf{H}^T\mathbf{HA})^{-1}\mathbf{A}^T\mathbf{H}^T\|\,\|\mathbf{HA}\|\frac{\|\mathbf{H}\hat{\boldsymbol{\epsilon}}\|}{\|\mathbf{Hb}\|}. \tag{2.5-16}$$

The quotients $\|\Delta\mathbf{x}\|/\|\mathbf{x}\|$ and $\|\mathbf{H}\hat{\boldsymbol{\epsilon}}\|/\|\mathbf{Hb}\|$ are measures of the relative errors in $\hat{\mathbf{x}}$ and $\hat{\mathbf{b}}$, respectively. Thus the inequality (2.5-16) gives an upper bound for the relative error in $\hat{\mathbf{x}}$ in terms of the relative error in $\hat{\mathbf{b}}$. It is a sharp upper bound, for it is possible to construct problems such that the equality holds exactly. Although it holds in general for any consistent vector and matrix norms, we will be primarily interested in the *Euclidean vector norm* and the consistent *spectral matrix norm,* which are defined, in the real case, by

$$\|\mathbf{z}\|_2 \equiv (\mathbf{z}^T\mathbf{z}) = (\text{length of vector } \mathbf{z}), \tag{2.5-17}$$

and

$$\begin{aligned}\|\mathbf{B}\|_2 = \max \xi_i(\mathbf{B}) &= (\text{maximum singular value of } \mathbf{B}) \\ &= (\text{maximum eigenvalue of } \mathbf{B}^T\mathbf{B})^{1/2}.\end{aligned} \tag{2.5-18}$$

These definitions have the same form in the complex case, but conjugate transposes are used rather than ordinary transposes. Using these norms, (2.5-16) becomes

$$\frac{\|\Delta\mathbf{x}\|_2}{\|\mathbf{x}\|_2} \leqslant \|(\mathbf{A}^T\mathbf{H}^T\mathbf{HA})^{-1}\mathbf{A}^T\mathbf{H}^T\|_2\,\|\mathbf{HA}\|_2\frac{\|\mathbf{H}\hat{\boldsymbol{\epsilon}}\|_2}{\|\mathbf{Hb}\|_2}. \tag{2.5-19}$$

Another matrix norm that is consistent with the Euclidean vector norm is the *Euclidean matrix norm*

$$\|\mathbf{B}\|_E \equiv \left(\sum_{i=1}^{m}\sum_{j=1}^{n} B_{ij}^2\right)^{1/2} = [\operatorname{trace}(\mathbf{B}^T\mathbf{B})]^{1/2}; \tag{2.5-20}$$

so (2.5-16) can also be written

$$\frac{\|\Delta\mathbf{x}\|_2}{\|\mathbf{x}\|_2} \leqslant \|(\mathbf{A}^T\mathbf{H}^T\mathbf{HA})^{-1}\mathbf{A}^T\mathbf{H}^T\|_E\,\|\mathbf{HA}\|_E\frac{\|\mathbf{H}\hat{\boldsymbol{\epsilon}}\|_2}{\|\mathbf{Hb}\|_2}. \tag{2.5-21}$$

The quantities $\|\mathbf{H}\hat{\boldsymbol{\epsilon}}\|_2$ and $\|\mathbf{Hb}\|_2$ are the Euclidean vector norms of the vectors $\mathbf{H}\hat{\boldsymbol{\epsilon}}$ and $\mathbf{Hb}$, but they can also be regarded as norms of the vectors $\hat{\boldsymbol{\epsilon}}$ and $\mathbf{b}$; that is,

$$\|\mathbf{H}\hat{\boldsymbol{\epsilon}}\|_2 = \|\hat{\boldsymbol{\epsilon}}\|_{\mathbf{S}^{-1}} \quad \text{and} \quad \|\mathbf{Hb}\|_2 = \|\mathbf{b}\|_{\mathbf{S}^{-1}}, \tag{2.5-22}$$

where in general the $\mathbf{S}^{-1}$ norm is defined by

$$\|\mathbf{z}\|_{\mathbf{S}^{-1}} \equiv (\mathbf{z}^T \mathbf{S}^{-1} \mathbf{z})^{1/2} = (\mathbf{z}^T \mathbf{H}^T \mathbf{H}\mathbf{z})^{1/2} = \|\mathbf{H}\mathbf{z}\|_2. \tag{2.5-23}$$

Thus, the inequality (2.5-19) can be written

$$\frac{\|\Delta\mathbf{x}\|_2}{\|\mathbf{x}\|_2} \leqslant \|(\mathbf{A}^T\mathbf{H}^T\mathbf{HA})^{-1}\mathbf{A}^T\mathbf{H}^T\|_2 \, \|\mathbf{HA}\|_2 \frac{\|\hat{\boldsymbol{\epsilon}}\|_{\mathbf{S}^{-1}}}{\|\mathbf{b}\|_{\mathbf{S}^{-1}}}. \tag{2.5-24}$$

We now consider the factor $\|(\mathbf{A}^T\mathbf{H}^T\mathbf{HA})^{-1}\mathbf{A}^T\mathbf{H}^T\|_2 \, \|\mathbf{HA}\|_2$, which relates the two relative errors in the bound (2.5-24). The $n \times m$ matrix $(\mathbf{A}^T\mathbf{H}^T\mathbf{HA})^{-1}\mathbf{A}^T\mathbf{H}^T$ is a left inverse for the matrix $\mathbf{HA}$ and is called the *generalized inverse* of $\mathbf{HA}$. More generally, for any $m \times n$ matrix $\mathbf{B}$ with $m \geqslant n$ and rank $(\mathbf{B}) = n$, the generalized inverse of $\mathbf{B}$ is the $n \times m$ matrix defined by

$$\mathbf{B}^\dagger = (\mathbf{B}^T\mathbf{B})^{-1}\mathbf{B}^T \tag{2.5-25}$$

where the nonsingularity of $\mathbf{B}^T\mathbf{B}$ follows from the linear independence of the columns of $\mathbf{B}$. Clearly $\mathbf{B}^\dagger$ is a left inverse of $\mathbf{B}$. Similarly, if $\mathbf{B}$ is an $m \times n$ matrix with $m \leqslant n$ and rank $(\mathbf{B}) = m$, then its generalized inverse is

$$\mathbf{B}^\dagger = \mathbf{B}^T(\mathbf{B}\mathbf{B}^T)^{-1} \tag{2.5-26}$$

and $\mathbf{B}^\dagger$ is a right inverse for $\mathbf{B}$. In the case where $\mathbf{B}$ is an $n \times n$ matrix with rank $(\mathbf{B}) = n$ (i.e., $\mathbf{B}$ is square and nonsingular), the generalized inverse $\mathbf{B}^\dagger$ is just the ordinary inverse $\mathbf{B}^{-1}$. Since $\mathbf{H}$ is an $m \times m$ nonsingular matrix, and $\mathbf{A}$ is an $m \times n$ matrix with linearly independent columns, it follows that $\mathbf{HA}$ is an $m \times n$ matrix with linearly independent columns, so the generalized inverse of $\mathbf{HA}$ is

$$(\mathbf{HA})^\dagger = [(\mathbf{HA})^T(\mathbf{HA})]^{-1}(\mathbf{HA})^T = (\mathbf{A}^T\mathbf{H}^T\mathbf{HA})^{-1}\mathbf{A}^T\mathbf{H}^T, \tag{2.5-27}$$

and thus the bound (2.5-24) can be written

$$\frac{\|\Delta\mathbf{x}\|_2}{\|\mathbf{x}\|_2} \leqslant \|(\mathbf{HA})^\dagger\|_2 \, \|\mathbf{HA}\|_2 \frac{\|\hat{\boldsymbol{\epsilon}}\|_{\mathbf{S}^{-1}}}{\|\mathbf{b}\|_{\mathbf{S}^{-1}}}. \tag{2.5-28}$$

The factor $\|(\mathbf{HA})^\dagger\|_2 \, \|\mathbf{HA}\|_2$ is called the *condition number* of the system of normal equations. If its value is small, a small relative change in $\hat{\mathbf{b}}$ cannot produce a very large relative change in $\hat{\mathbf{x}}$, and thus the normal equations are *well conditioned.* But if it has a large value, small relative changes in $\hat{\mathbf{b}}$ may produce large relative changes in $\hat{\mathbf{x}}$, and the system is said to be *poorly conditioned* or *ill conditioned.*

Since the spectral norm of a matrix is equal to its largest singular value, we can express the condition number $\|(\mathbf{HA})^\dagger\|_2 \|\mathbf{HA}\|_2$ in terms of the singular values of **HA**. We already have

$$\|\mathbf{HA}\|_2 = \omega_1, \tag{2.5-29}$$

and to find the largest singular value of the matrix $(\mathbf{HA})^\dagger$ we substitute the factorization (2.5-10) into Eq. (2.5-27) to get

$$(\mathbf{HA})^\dagger = \mathbf{P}(\boldsymbol{\Omega}^{-1}, \mathbf{0})\,\mathbf{L}^T. \tag{2.5-30}$$

Since $(\mathbf{HA})^\dagger$ is an $n \times m$ matrix with $m \geqslant n$, its singular values are the positive square roots of the eigenvalues of the symmetric $n \times n$ matrix

$$(\mathbf{HA})^\dagger\ [(\mathbf{HA})^\dagger]^T = \mathbf{P}(\boldsymbol{\Omega}^{-1})^2\mathbf{P}^T.$$

The eigenvalues of this matrix are just the diagonal elements of

$$(\boldsymbol{\Omega}^{-1})^2 = \operatorname{diag}\left(\frac{1}{\omega_1^2}, \frac{1}{\omega_2^2}, \ldots, \frac{1}{\omega_n^2}\right),$$

so the singular values of $(\mathbf{HA})^\dagger$ are just

$$\frac{1}{\omega_1} \leqslant \frac{1}{\omega_2} \leqslant \cdots \leqslant \frac{1}{\omega_n},$$

that is, the reciprocals of the singular values of **HA**. The largest singular value of $(\mathbf{HA})^\dagger$ is $1/\omega_n$, so

$$\|(\mathbf{HA})^\dagger\|_2 = \frac{1}{\omega_n} \tag{2.5-31}$$

and the condition number is

$$\|\mathbf{HA}\|_2\, \|(\mathbf{HA})^\dagger\|_2 = \frac{\omega_1}{\omega_n}. \tag{2.5-32}$$

Therefore the bound (2.5-28) can be written

$$\frac{\|\Delta\mathbf{x}\|_2}{\|\mathbf{x}\|_2} \leqslant \frac{\omega_1}{\omega_n} \frac{\|\hat{\boldsymbol{\epsilon}}\|_{\mathbf{S}^{-1}}}{\|\mathbf{b}\|_{\mathbf{S}^{-1}}}; \tag{2.5-33}$$

it is the ratio of the largest to the smallest singular value that determines whether or not the system is poorly conditioned. It might appear from Eq. (2.5-12) that, if all the singular values are very small, a small error $\hat{\boldsymbol{\epsilon}}$ might produce a very large error $\Delta\mathbf{x}$; but it is apparent from the bound above that no great magnification of the relative error can occur if the singular values do not vary too much in magnitude.

The condition number can also be expressed as the ratio of the longest to the shortest principal axis of the κ-ellipsoid. By Eq. (2.4-13) the length of the longest principal axis is

$$\Gamma_n = \frac{2\kappa}{\omega_n}$$

and the length of the shortest principal axis is

$$\Gamma_1 = \frac{2\kappa}{\omega_1}.$$

Therefore the ratio of the two is

$$\frac{\Gamma_n}{\Gamma_1} = \frac{2\kappa/\omega_n}{2\kappa/\omega_1} = \frac{\omega_1}{\omega_n}.$$

This means that the normal equations are ill conditioned only if at least one of the principal axes of the κ-ellipsoid is either much larger or much smaller than the others. Figure 2.4 illustrates, for $n = 2$, the difference in the κ-ellipsoids for a well-conditioned and a poorly conditioned problem.

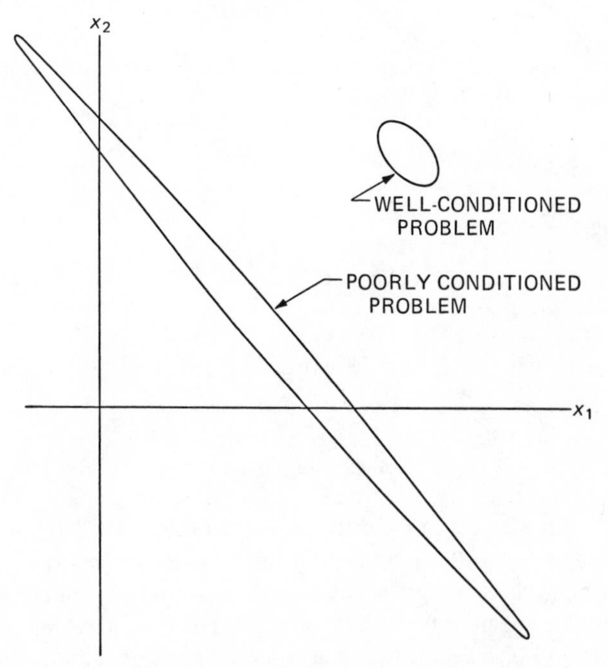

Figure 2.4. κ-Ellipsoids for a well-conditioned and a poorly conditioned problem.

When the normal equations are poorly conditioned, the estimator $\hat{\mathbf{x}}$ is a highly unstable function of the observed value $\hat{\mathbf{b}}$. Two slightly different values of the observation vector $\hat{\mathbf{b}}$ are likely to produce vastly differing values of $\hat{\mathbf{x}}$. In Section 2.3 we made the assumption that $\hat{\mathbf{b}}$ has a joint normal distribution, so its probability density function can be written

$$g(\hat{\mathbf{b}}) = [(2\pi)^m \, |\mathbf{S}|]^{-1/2} \exp\,[-\tfrac{1}{2}(\hat{\mathbf{b}} - \mathbf{b})^T \mathbf{S}^{-1} (\hat{\mathbf{b}} - \mathbf{b})] \qquad (2.5\text{-}34)$$

where $\mathbf{b} = \mathbf{Ax}$ is the true, but unknown, value of $\hat{\mathbf{b}}$. If γ^2 is any positive constant, then the equation

$$(\hat{\mathbf{b}} - \mathbf{b})^T \mathbf{S}^{-1} (\hat{\mathbf{b}} - \mathbf{b}) = \gamma^2 \qquad (2.5\text{-}35)$$

defines an ellipsoid in the $\hat{b}$-space. This ellipsoid is centered at the point $\mathbf{b}$, which is the mean of the $\hat{\mathbf{b}}$ distribution, and the probability density function $g(\hat{\mathbf{b}})$ has the same constant value at every point on its surface. The shape and orientation of the ellipsoid are determined by the eigenvectors and eigenvalues of $\mathbf{S}^{-1}$, and its size is determined by the value of γ^2. For larger and larger values of γ^2, the ellipsoid becomes larger and larger, the associated value of the probability distribution function becomes smaller and smaller, and the amount of probability density enclosed inside the ellipsoid becomes greater and greater (approaching 1 as γ^2 approaches ∞).

Suppose we pick some value for γ^2 that is large enough to enclose most of the probability density in the interior of the ellipsoid; that is, in the region β defined by

$$\beta\colon (\hat{\mathbf{b}} - \mathbf{b})^T \mathbf{S}^{-1}(\hat{\mathbf{b}} - \mathbf{b}) \leqslant \gamma^2. \qquad (2.5\text{-}36)$$

We assume that the concentration of the probability density in the region around the true value $\mathbf{b}$ is great enough so that the dimensions of the ellipsoid are small in comparison to the length of the vector $\mathbf{b}$. This means that any vector $\hat{\mathbf{b}}$ that lies in the region β has a small relative error $\| \hat{\boldsymbol{\epsilon}} \|_2 / \| \mathbf{b} \|_2$. Now suppose that for each vector $\hat{\mathbf{b}}$ in the region β, the corresponding estimate $\hat{\mathbf{x}}$ is computed. It is fairly easy to see that the resulting set of estimates will form the interior and boundary of an ellipsoid in the $\hat{\mathbf{x}}$-space. From Eqs. (2.5-1), (2.5-4), and (2.5-27) it follows that the estimate $\hat{\mathbf{x}}$ can be written

$$\hat{\mathbf{x}} = (\mathbf{HA})^\dagger \, \mathbf{H}\hat{\mathbf{b}}, \qquad (2.5\text{-}37)$$

and by Eq. (2.5-30) it follows that

$$\hat{\mathbf{x}} = \mathbf{P}(\boldsymbol{\Omega}^{-1}, \mathbf{0}) \, \mathbf{L}^T \mathbf{H}\hat{\mathbf{b}}. \qquad (2.5\text{-}38)$$

Now, the $\hat{\mathbf{b}}$-vectors are distributed in the region β defined by Eq. (2.5-36), which can also be written

$$(\hat{\mathbf{b}} - \mathbf{b})^T \mathbf{H}^T \mathbf{H}(\hat{\mathbf{b}} - \mathbf{b}) \leqslant \gamma^2 \qquad (2.5\text{-}39)$$

or simply

$$(\hat{\mathbf{w}} - \mathbf{Hb})^T (\hat{\mathbf{w}} - \mathbf{Hb}) \leqslant \gamma^2 \tag{2.5-40}$$

where $\hat{\mathbf{w}}$ is the new variable defined by

$$\hat{\mathbf{w}} = \mathbf{H}\hat{\mathbf{b}}. \tag{2.5-41}$$

Equation (2.5-40) is the equation of the interior and surface of an m-dimensional hypersphere in the w-space. Thus, the vectors $\mathbf{H}\hat{\mathbf{b}}$ are distributed inside a spheroid (see Fig. 2.5). Now consider the distribution of the vectors

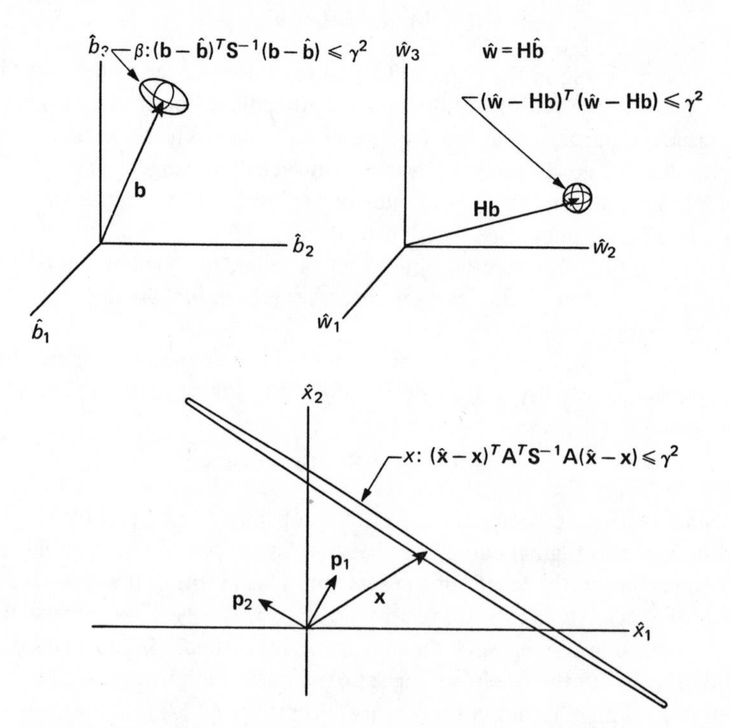

Figure 2.5. An ill-conditioned problem with $m = 3$ and $n = 2$.

$\mathbf{L}^T\mathbf{H}\hat{\mathbf{b}}$. The matrix $\mathbf{L}$ is orthogonal, so premultiplication by the matrix $\mathbf{L}^T$ simply rotates all the vectors $\mathbf{H}\hat{\mathbf{b}}$ through the same angle. Therefore the vectors $\mathbf{L}^T\mathbf{H}\hat{\mathbf{b}}$ are also distributed inside an m-dimensional hypersphere. Now consider the effect of multiplying all the vectors in this spheroid by the matrix $(\mathbf{\Omega}^{-1}, \mathbf{0})$. Because of the $n \times (m - n)$ zero submatrix on the right, one effect of this transformation is to project the whole space orthogonally into the n-dimensional subspace corresponding to the first n coordinates. The orthogonal projection of the m-dimensional hypersphere into the n-dimensional subspace yields an n-dimensional hypersphere. The diagonal submatrix $\mathbf{\Omega}^{-1}$ transforms the vectors in the n-dimensional subspace, and if these vectors are regarded as n-vectors, then the effect of the multiplication by $\mathbf{\Omega}^{-1}$ is to stretch each of the n-components by an amount proportional to the corresponding diagonal element. Thus, unless $\omega_1 = \omega_2 = \cdots = \omega_n$, spheroids in this n-space will be stretched into ellipsoids and the vectors $(\mathbf{\Omega}^{-1}, \mathbf{0})\mathbf{L}^T\mathbf{H}\hat{\mathbf{b}}$ will be distributed inside and on the surface of an n-dimensional ellipsoid. The matrix $\mathbf{P}$ is orthogonal, so its effect on these vectors will be to rotate them all through the same angle. Therefore, the vectors $\hat{\mathbf{x}} = \mathbf{P}(\mathbf{\Omega}^{-1}, \mathbf{0})\mathbf{L}^T\mathbf{H}\hat{\mathbf{b}}$ will be distributed inside and on the surface of an ellipsoid X in the x-space.

To determine the equation of the ellipsoidal region X, we begin with Eq. (2.5-39), which, since $\mathbf{L}$ is an orthogonal matrix, can be written

$$(\hat{\mathbf{b}} - \mathbf{b})^T \mathbf{H}^T \mathbf{L}\mathbf{L}^T \mathbf{H}(\hat{\mathbf{b}} - \mathbf{b}) \leqslant \gamma^2$$

or

$$(\mathbf{L}^T \mathbf{H}\hat{\mathbf{b}} - \mathbf{L}^T \mathbf{H}\mathbf{b})^T(\mathbf{L}^T \mathbf{H}\hat{\mathbf{b}} - \mathbf{L}^T \mathbf{H}\mathbf{b}) \leqslant \gamma^2.$$

This last equation defines the region inside and on a hypersphere in m-space, and the equation of the spheroidal region that is its orthogonal projection into the space of the first n-components is

$$(\mathbf{L}^T \mathbf{H}\hat{\mathbf{b}} - \mathbf{L}^T \mathbf{H}\mathbf{b})^T \begin{pmatrix} \mathbf{I}_n & \mathbf{0} \\ \mathbf{0} & \mathbf{0} \end{pmatrix} (\mathbf{L}^T \mathbf{H}\hat{\mathbf{b}} - \mathbf{L}^T \mathbf{H}\mathbf{b}) \leqslant \gamma^2 \qquad (2.5\text{-}42)$$

where the matrix in the middle is $m \times m$ with the nth-order identity matrix as its upper left-hand $n \times n$ submatrix. This matrix can be expressed

$$\begin{pmatrix} \mathbf{I}_n & \mathbf{0} \\ \mathbf{0} & \mathbf{0} \end{pmatrix} = \begin{pmatrix} \mathbf{\Omega}^{-1} \\ \mathbf{0} \end{pmatrix} \mathbf{\Omega}^2 (\mathbf{\Omega}^{-1}, \mathbf{0});$$

if this expression is substituted into Eq. (2.5-42), the result can be written

$$[(\mathbf{\Omega}^{-1}, \mathbf{0})\mathbf{L}^T \mathbf{H}\hat{\mathbf{b}} - (\mathbf{\Omega}^{-1}, \mathbf{0})\mathbf{L}^T \mathbf{H}\mathbf{b}]^T \mathbf{\Omega}^2 [(\mathbf{\Omega}^{-1}, \mathbf{0})\mathbf{L}^T \mathbf{H}\hat{\mathbf{b}} - (\mathbf{\Omega}^{-1}, \mathbf{0})\mathbf{L}^T \mathbf{H}\mathbf{b}] \leqslant \gamma^2.$$

Since $\mathbf{P}$ is an orthogonal matrix, the last equation can also be written

$$[\mathbf{P}(\mathbf{\Omega}^{-1}, \mathbf{0})\mathbf{L}^T \mathbf{H}\hat{\mathbf{b}} - \mathbf{P}(\mathbf{\Omega}^{-1}, \mathbf{0})\mathbf{L}^T \mathbf{H}\mathbf{b}]^T \mathbf{P}\mathbf{\Omega}^2 \mathbf{P}^T [\mathbf{P}(\mathbf{\Omega}^{-1}, \mathbf{0})\mathbf{L}^T \mathbf{H}\hat{\mathbf{b}} - \mathbf{P}(\mathbf{\Omega}^{-1}, \mathbf{0})\mathbf{L}^T \mathbf{H}\mathbf{b}] \leqslant \gamma^2,$$

which, by Eqs, (2.4-9) and (2.5-38), is the same as

$$(\hat{\mathbf{x}} - \mathbf{x})^T \mathbf{A}^T \mathbf{S}^{-1} \mathbf{A}(\hat{\mathbf{x}} - \mathbf{x}) \leqslant \gamma^2. \qquad (2.5\text{-}43)$$

This is the equation of the region X. A comparison with Eq. (2.4-2) reveals that this region has the same shape and orientation as the κ-ellipsoids. The principal axes have the same directions as the eigenvectors $\mathbf{p}_1, \mathbf{p}_2, \ldots, \mathbf{p}_n$, and their lengths $2\Upsilon/\omega_1, 2\Upsilon/\omega_2, \ldots, 2\Upsilon/\omega_n$ are inversely proportional to the corresponding singular values. This distribution is exactly what we might have expected on the basis of Eq. (2.5-12) and the discussion that followed it. The system is ill conditioned only if the principal axis in the direction of $\mathbf{p}_n$ is very much longer than the axis in the direction of $\mathbf{p}_1$. Figure 2.5 illustrates the regions β and X for an ill-conditioned problem with $m = 3$ and $n = 2$.

2.6 The Generalized Inverse

In the last section we saw that the condition of the normal equations depends on the ratio ω_1/ω_n of the largest to the smallest singular value of the matrix **HA**. The larger this ratio is, the more ill conditioned is the system of normal equations. Although this ratio can become quite large, it will always have a finite value for any problem that satisfies the assumptions of the classical linear regression model. One of these assumptions is that the matrix **A** has full rank n. This assumption guarantees that there is a unique solution vector $\hat{\mathbf{x}}$ and that point estimates and interval estimates can be found for any linear function $\phi = \mathbf{c}^T\mathbf{x}$. If this assumption is relaxed, that is, if rank $(\mathbf{A}) < n$, then one or more of the singular values of **HA** will be equal to zero; in particular, $\omega_n = 0$, so that the condition number ω_1/ω_n is infinite and there is no best estimator $\hat{\mathbf{x}}$ of the unknown vector $\mathbf{x}$. Many linear functions $\phi = \mathbf{c}^T\mathbf{x}$ will not be estimable in this case, but the linear estimation procedure does not completely break down, for many other linear functions will be estimable. In this section we drop the assumption that rank $(\mathbf{A}) = n$ and determine which functions are estimable and the estimators for them.

In order to discuss estimation for problems with rank-deficient matrices, we need the idea of the *generalized inverse* of a matrix. In the last section we discussed generalized inverses for rectangular matrices with full rank, but the concept of the generalized inverse can be extended to matrices with less than full rank. If **B** is any $m \times n$ real matrix, then it can be shown [9] that there exists a unique real $n \times m$ matrix $\mathbf{B}^\dagger$ that satisfies the following four conditions:

$$(\mathbf{B}^\dagger\mathbf{B})^T = \mathbf{B}^\dagger\mathbf{B}, \tag{2.6-1}$$

$$(\mathbf{B}\mathbf{B}^\dagger)^T = \mathbf{B}\mathbf{B}^\dagger, \tag{2.6-2}$$

$$\mathbf{B}\mathbf{B}^\dagger\mathbf{B} = \mathbf{B}, \tag{2.6-3}$$

$$\mathbf{B}^\dagger\mathbf{B}\mathbf{B}^\dagger = \mathbf{B}^\dagger. \tag{2.6-4}$$

The matrix $\mathbf{B}^\dagger$ is called the generalized inverse of the matrix **B**. If **B** is square and nonsingular, then $\mathbf{B}^\dagger = \mathbf{B}^{-1}$, and if **B** is rectangular and of full rank, then $\mathbf{B}^\dagger$ is given by Eq. (2.5-25) or (2.5-26), depending on which is greater, the number of columns or the number of rows. The generalized inverse also has the following properties.

$$(\mathbf{B}^\dagger)^\dagger = \mathbf{B}, \tag{2.6-5}$$

$$(\mathbf{B}^T)^\dagger = (\mathbf{B}^\dagger)^T, \tag{2.6-6}$$

$$\mathbf{B}^\dagger \mathbf{B}\mathbf{B}^T = \mathbf{B}^T = \mathbf{B}^T \mathbf{B}\mathbf{B}^\dagger, \tag{2.6-7}$$

$$\mathbf{B}^\dagger(\mathbf{B}^\dagger)^T \mathbf{B}^T = \mathbf{B}^\dagger = \mathbf{B}^T(\mathbf{B}^\dagger)^T \mathbf{B}^\dagger, \tag{2.6-8}$$

$$(\mathbf{B}^T \mathbf{B})^\dagger = \mathbf{B}^\dagger(\mathbf{B}^\dagger)^T, \tag{2.6-9}$$

$$\mathbf{U}, \mathbf{V} \text{ orthogonal} \Rightarrow (\mathbf{U}\mathbf{B}\mathbf{V})^\dagger = \mathbf{V}^T \mathbf{B}^\dagger \mathbf{U}^T, \tag{2.6-10}$$

$$\mathbf{B}^\dagger = (\mathbf{B}^T \mathbf{B})^\dagger \mathbf{B}^T. \tag{2.6-11}$$

The foregoing definition and properties hold for complex matrices also, if the conjugate transpose is used in place of the ordinary transpose.

We now consider the case where the matrix **B** has less than full rank. We assume for the sake of definiteness that $m > n$ and rank $(\mathbf{B}) = \mathrm{p} < n$. We saw in the last section [Eqs. (2.5-6) and (2.5-7)] that **B** can be factored

$$\mathbf{B} = \mathbf{L} \begin{bmatrix} \mathbf{D} \\ \mathbf{0} \end{bmatrix} \mathbf{R}^T \tag{2.6-12}$$

where **L** and **R** are orthogonal and **D** is an $n \times n$ diagonal matrix whose diagonal elements are the singular values of **B**. Therefore, by (2.6-10) the generalized inverse of **B** is given by

$$\mathbf{B}^\dagger = \mathbf{R} \begin{bmatrix} \mathbf{D} \\ \mathbf{0} \end{bmatrix}^\dagger \mathbf{L}^T.$$

It is easy to verify, by substitution into Eqs. (2.6-1, 2, 3, 4), that

$$\begin{bmatrix} \mathbf{D} \\ \mathbf{0} \end{bmatrix}^\dagger = [\mathbf{D}^\dagger, \mathbf{0}],$$

so that

$$\mathbf{B}^\dagger = \mathbf{R}[\mathbf{D}^\dagger, \mathbf{0}]\mathbf{L}^T. \tag{2.6-13}$$

Now $n-p$ of the singular values of $\mathbf{B}$ are equal to zero, so $\mathbf{D}$ is an $n \times n$ matrix of the form

$$\mathbf{D} = \operatorname{diag}(\xi_1, \xi_2, \ldots, \xi_p, 0, \ldots, 0) \tag{2.6-14}$$

where $\xi_1, \xi_2, \ldots, \xi_p$ are the nonzero singular values. It is easy to verify that

$$\mathbf{D}^\dagger = \operatorname{diag}(1/\xi_1, 1/\xi_2, \ldots, 1/\xi_p, 0, \ldots, 0). \tag{2.6-15}$$

In the case where $m < n$ and rank $(\mathbf{B}) = p < m$, the factorization for $\mathbf{B}$ is

$$\mathbf{B} = \mathbf{L}[\mathbf{D}, \mathbf{0}]\,\mathbf{R}^T \tag{2.6-16}$$

and $\mathbf{B}^\dagger$ is

$$\mathbf{B}^\dagger = \mathbf{R}\begin{bmatrix} \mathbf{D}^\dagger \\ \mathbf{0} \end{bmatrix}\mathbf{L}^T. \tag{2.6-17}$$

where $\mathbf{D}$ is an $m \times m$ matrix like the one in Eq. (2.6-14).

We now consider the general system of linear equations

$$\mathbf{Bz} = \mathbf{d} \tag{2.6-18}$$

where $\mathbf{B}$ is an $m \times n$ matrix and $\mathbf{d}$ is an m-vector. A necessary and sufficient condition for this equation to be solvable is that

$$\mathbf{BB}^\dagger\mathbf{d} = \mathbf{d}. \tag{2.6-19}$$

To see this, we note that if (2.6-19) holds, then $\mathbf{B}^\dagger\mathbf{d}$ is a solution of (2.6-18), and if (2.6-18) is true for some vector $\mathbf{z}$, then

$$\mathbf{BB}^\dagger\,\mathbf{Bz} = \mathbf{BB}^\dagger\,\mathbf{d},$$

which by (2.6-3) is the same as

$$\mathbf{Bz} = \mathbf{BB}^\dagger\mathbf{d} \qquad \text{or simply} \qquad \mathbf{d} = \mathbf{BB}^\dagger\mathbf{d}.$$

The condition expressed by (2.6-19) can be regarded as a *consistency condition* and the system (2.6-18) has solutions if and only if it is satisfied; that is, if and only if the system is consistent. When the system is consistent it may have a unique solution or many solutions, depending on the dimensions and rank of the matrix $\mathbf{B}$. Table 2.2 lists the possibilities.

To determine the general solution of the system (2.6-18), we first consider the homogeneous system

$$\mathbf{Bz} = \mathbf{0}. \tag{2.6-20}$$

For any n-vector v, the vector

$$\mathbf{z} = (\mathbf{I} - \mathbf{B}^\dagger\,\mathbf{B})\,\mathbf{v} \tag{2.6-21}$$

is a solution to the homogeneous system, since

$$\mathbf{Bz} = \mathbf{B}(\mathbf{I} - \mathbf{B}^{\dagger}\mathbf{B})\mathbf{v} = (\mathbf{B} - \mathbf{B}\mathbf{B}^{\dagger}\mathbf{B})\mathbf{v} = \mathbf{0};$$

any solution of (2.6-20) can be expressed in the form (2.6-21) because $\mathbf{Bz} = \mathbf{0}$ implies that

$$\mathbf{B}^{\dagger}\mathbf{Bz} = \mathbf{0},$$

TABLE 2.2 Possible consistent systems of linear equations

$m = n$	Rank (B) $= n$, unique solution $\mathbf{z} = \mathbf{B}^{-1}\mathbf{d}$
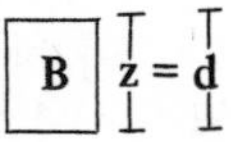 B z = d	Rank (B) $< n$, many solutions
$m > n$	Rank (B) $= n$, unique solution $\mathbf{z} = (\mathbf{B}^T\mathbf{B})^{-1}\mathbf{B}^T\mathbf{d}$
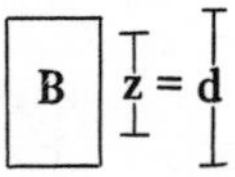 B z = d	Rank (B) $< n$, many solutions
$m < n$	Many solutions
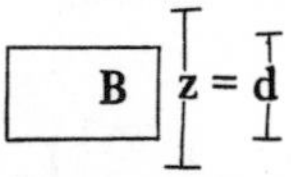 B z = d	

from which it follows that

$$\mathbf{z} = \mathbf{z} - \mathbf{B}^{\dagger}\mathbf{Bz} = (\mathbf{I} - \mathbf{B}^{\dagger}\mathbf{B})\mathbf{z}.$$

Thus the expression (2.6-21), where $\mathbf{v}$ is an arbitrary n-vector, is the general solution of the homogeneous equation (2.6-20). Now consider the nonhomogeneous equation (2.6-18). We have seen that this equation is solvable if and only if the consistency condition (2.6-19) is satisfied, and if this condition is satisfied, then a particular solution is

$$\mathbf{z} = \mathbf{B}^{\dagger}\mathbf{d}. \tag{2.6-22}$$

Combining the general solution (2.6-21) and the particular solution (2.6-22) gives the general solution of the nonhomogeneous system (2.6-18), that is,

$$\mathbf{z} = \mathbf{B}^{\dagger}\mathbf{d} + (\mathbf{I} - \mathbf{B}^{\dagger}\mathbf{B})\mathbf{v} \tag{2.6-23}$$

where $\mathbf{v}$ is an arbitrary n-vector.

The general solution (2.6-23) has an interesting geometrical interpretation.

Consider first the set of all solutions to the homogeneous equation; that is, the set

$$S = \{\mathbf{z} | \mathbf{z} = (\mathbf{I} - \mathbf{B}^\dagger \mathbf{B})\mathbf{v} \text{ for some } \mathbf{v} \text{ in } E_n\}, \tag{2.6-24}$$

where E_n denotes the Euclidean n-space. If $\mathbf{z}_1$ and $\mathbf{z}_2$ are any two vectors in this set and ζ_1 and ζ_2 are any two scalars, then

$$\mathbf{z}_1 = (\mathbf{I} - \mathbf{B}^\dagger \mathbf{B})\mathbf{v}_1, \qquad \mathbf{z}_2 = (\mathbf{I} - \mathbf{B}^\dagger \mathbf{B})\mathbf{v}_2$$

for some $\mathbf{v}_1$ and $\mathbf{v}_2$ in E_n, and

$$\zeta_1 \mathbf{z}_1 + \zeta_2 \mathbf{z}_2 = (\mathbf{I} - \mathbf{B}^\dagger \mathbf{B})(\zeta_1 \mathbf{v}_1 + \zeta_2 \mathbf{v}_2).$$

Now $(\zeta_1 \mathbf{v}_1 + \zeta_2 \mathbf{v}_2) \in E_n$, so the vector $\zeta_1 \mathbf{z}_1 + \zeta_2 \mathbf{z}_2$ must belong to the set S. This means that S is closed under the operations of vector addition and multiplication of a vector by a scalar, so S is a subspace of the Euclidean n-space. Since S is the set of all possible linear combinations of the columns of the matrix $(\mathbf{I} - \mathbf{B}^\dagger\mathbf{B})$, the dimension of S is equal to the rank of $(\mathbf{I} - \mathbf{B}^\dagger\mathbf{B})$. From Eqs. (2.6-12) and (2.6-13) [or (2.6-16) and (2.6-17)], we have

$$\mathbf{B}^\dagger \mathbf{B} = \mathbf{R} \begin{bmatrix} \mathbf{I}_p & \mathbf{0} \\ \mathbf{0} & \mathbf{0}_{n-p} \end{bmatrix} \mathbf{R}^T$$

where p is the rank of $\mathbf{B}$, $\mathbf{I}_p$ is the pth-order identity matrix, and $\mathbf{0}_{n-p}$ is an $(n - p) \times (n - p)$ zero matrix. It follows then that

$$\mathbf{I} - \mathbf{B}^\dagger \mathbf{B} = \mathbf{I}_n - \mathbf{R} \begin{bmatrix} \mathbf{I}_p & \mathbf{0} \\ \mathbf{0} & \mathbf{0}_{n-p} \end{bmatrix} \mathbf{R}^T = \mathbf{R} \left\{ \mathbf{I}_n - \begin{bmatrix} \mathbf{I}_p & \mathbf{0} \\ \mathbf{0} & \mathbf{0}_{n-p} \end{bmatrix} \right\} \mathbf{R}^T,$$

or

$$\mathbf{I} - \mathbf{B}^\dagger \mathbf{B} = \mathbf{R} \begin{bmatrix} \mathbf{0}_p & \mathbf{0} \\ \mathbf{0} & \mathbf{I}_{n-p} \end{bmatrix} \mathbf{R}^T,$$

and the rank of this matrix is just $(n - p)$. Thus the dimension of the subspace S is equal to $(n - p)$.

Now consider the set of all solutions (2.6-23) of the nonhomogeneous system; that is, the set

$$M = \{\mathbf{z} | \mathbf{z} = \mathbf{B}^\dagger \mathbf{d} + (\mathbf{I} - \mathbf{B}^\dagger \mathbf{B})\mathbf{v} \text{ for some } \mathbf{v} \text{ in } E_n\}. \tag{2.6-25}$$

This is just the set that is obtained by adding the vector $\mathbf{B}^\dagger\mathbf{d}$ to every vector in the subspace S. Hence it is just a displaced or translated subspace or a *coset of the subspace S*. The matrix $\mathbf{B}^\dagger\mathbf{d}$ is orthogonal to S since for any vector $\mathbf{v}$

$$\begin{aligned} [(\mathbf{I} - \mathbf{B}^\dagger \mathbf{B})\mathbf{v}]^T \mathbf{B}^\dagger \mathbf{d} &= \mathbf{v}^T(\mathbf{I} - \mathbf{B}^\dagger \mathbf{B})^T \mathbf{B}^\dagger \mathbf{d} = \mathbf{v}^T(\mathbf{I} - \mathbf{B}^\dagger \mathbf{B})\mathbf{B}^\dagger \mathbf{d} \\ &= \mathbf{v}^T(\mathbf{B}^\dagger - \mathbf{B}^\dagger \mathbf{B}\mathbf{B}^\dagger)\mathbf{d} = 0. \end{aligned}$$

The relationship between the set M and the subspace S is shown in Fig. 2.6 for a system with $n = 2$ and $p = 1$. Any vector whose tip lies on the line M is a solution. It is clear from the figure that for this problem the solution vector with shortest length is the vector $\mathbf{B}^\dagger\mathbf{d}$. In fact, it is true in general that the shortest-length solution vector is $\mathbf{B}^\dagger\mathbf{d}$, since any solution can be expressed as the sum of $\mathbf{B}^\dagger\mathbf{d}$ and a vector orthogonal to $\mathbf{B}^\dagger\mathbf{d}$, and the sum of the two will have minimal length when the component orthogonal to $\mathbf{B}^\dagger\mathbf{d}$ is equal to zero.

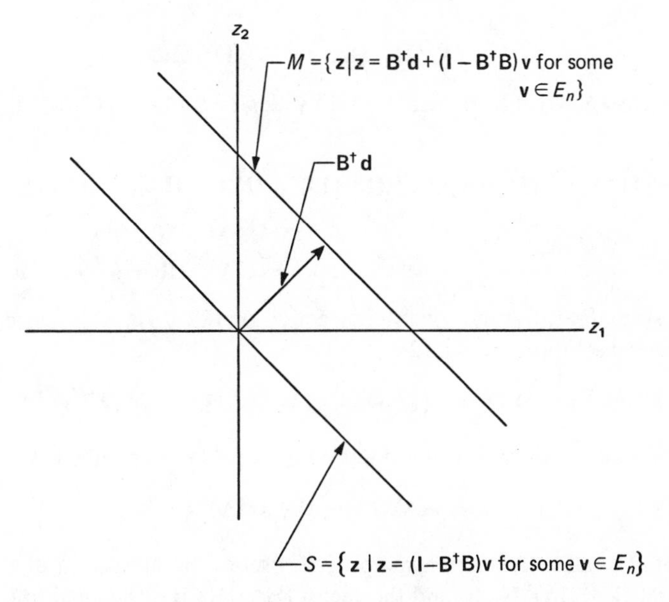

Figure 2.6. The coset of all solutions of a system with $n = 2$ and $p = 1$.

We now return to the modified linear regression model

$$\hat{\mathbf{b}} = \mathbf{A}\mathbf{x} + \hat{\boldsymbol{\epsilon}} \tag{2.6-26}$$

where $\mathbf{A}$ is a matrix with rank $p < n$. There is no unbiased linear estimator

$$\hat{\mathbf{x}} = \mathbf{U}^T\hat{\mathbf{b}},$$

for such an estimator would be unbiased only if

$$E(\hat{\mathbf{x}}) \equiv \mathbf{x} \qquad \text{or} \qquad \mathbf{U}^T\mathbf{A}\mathbf{x} \equiv \mathbf{x}$$

whatever the value of $\mathbf{x}$, and this is true only if the $n \times m$ matrix $\mathbf{U}$ satisfies

$$\mathbf{U}^T\mathbf{A} = \mathbf{I}_n.$$

No such matrix $\mathbf{U}$ exists, for if it did, then the matrix product $\mathbf{U}^T\mathbf{A}$ would have a greater rank than one of its factors. Thus there is no unbiased linear estimator for $\mathbf{x}$. It is, however, still possible to minimize the least squares quadratic form

$$\theta(\mathbf{z}) = (\hat{\mathbf{b}} - \mathbf{Az})^T \mathbf{S}^{-1}(\hat{\mathbf{b}} - \mathbf{Az}). \tag{2.6-27}$$

The normal equations are

$$(\mathbf{A}^T\mathbf{S}^{-1}\mathbf{A})\mathbf{z} = \mathbf{A}^T\mathbf{S}^{-1}\hat{\mathbf{b}}$$

or

$$(\mathbf{A}^T\mathbf{H}^T\mathbf{HA})\mathbf{z} = \mathbf{A}^T\mathbf{H}^T\mathbf{H}\hat{\mathbf{b}}. \tag{2.6-28}$$

The matrix $(\mathbf{A}^T\mathbf{H}^T\mathbf{HA})$ is singular, but by Eqs. (2.6-11) and (2.6-7) it follows that

$$\begin{aligned}(\mathbf{A}^T\mathbf{H}^T\mathbf{HA})(\mathbf{A}^T\mathbf{H}^T\mathbf{HA})^\dagger(\mathbf{A}^T\mathbf{H}^T\mathbf{H}\hat{\mathbf{b}}) &= (\mathbf{HA})^T\mathbf{HA}[(\mathbf{HA})^T\mathbf{HA}]^\dagger(\mathbf{HA})^T\mathbf{H}\hat{\mathbf{b}}\\ &= (\mathbf{HA})^T\mathbf{HA}(\mathbf{HA})^\dagger\mathbf{H}\hat{\mathbf{b}}\\ &= (\mathbf{HA})^T\mathbf{H}\hat{\mathbf{b}} = \mathbf{A}^T\mathbf{H}^T\mathbf{H}\hat{\mathbf{b}};\end{aligned}$$

so the system satisfies the consistency condition (2.6-19) and is hence solvable. The general solution, by Eq. (2.6-23), is

$$\mathbf{z} = (\mathbf{A}^T\mathbf{H}^T\mathbf{HA})^\dagger\mathbf{A}^T\mathbf{H}^T\mathbf{H}\hat{\mathbf{b}} + [\mathbf{I} - (\mathbf{A}^T\mathbf{H}^T\mathbf{HA})^\dagger(\mathbf{A}^T\mathbf{H}^T\mathbf{HA})]\mathbf{v}$$

where $\mathbf{v}$ is an arbitrary n-vector, and by Eq. (2.6-11) this expression reduces to

$$\mathbf{z} = (\mathbf{HA})^\dagger\mathbf{H}\hat{\mathbf{b}} + [\mathbf{I} - (\mathbf{HA})^\dagger\mathbf{HA}]\mathbf{v}. \tag{2.6-29}$$

The set M of all such solution vectors is a coset of the subspace S of all vectors of the form $[\mathbf{I} - (\mathbf{HA})^\dagger\mathbf{HA}]\mathbf{v}$, and the vector $(\mathbf{HA})^\dagger\mathbf{H}\hat{\mathbf{b}}$ is orthogonal to this subspace. The quadratic form (2.6-27) attains its minimum value r_0 at every point in the set M, and the vector $(\mathbf{HA})^\dagger\mathbf{H}\hat{\mathbf{b}}$ is the shortest vector that yields this minimum value.

Figure 2.7 shows the set M and residual surface for a problem with $n = 2$ and $p = 1$. The residual surface is an infinitely long quadratic trough and the coset M is the projection onto the z-plane of the bottom of that trough. The κ-ellipsoids, which are obtained by cutting the residual surface with planes parallel to the z-plane and projecting the intersection onto the z-plane, have as one of their principal axes the coset M. Thus, each of the κ-ellipsoids consists of a pair of lines parallel to the coset M, one on each side of M, and equally distant from M. In the general case, when the matrix $\mathbf{A}$ has full rank, the principal axes have the same directions as the eigenvectors $\mathbf{p}_1, \mathbf{p}_2, \ldots, \mathbf{p}_n$ of $\mathbf{A}^T\mathbf{H}^T\mathbf{HA}$ and lengths $2\kappa/\omega_1, 2\kappa/\omega_2, \ldots, 2\kappa/\omega_n$ where ω_j is the singular value of $\mathbf{HA}$ corresponding to the eigenvector $\mathbf{p}_j$. But we are assuming here that the matrix $\mathbf{A}$ has rank $p < n$, so that the κ-ellipsoid will be infinitely wide in any direction parallel to the subspace spanned by the eigenvectors corresponding to the zero singular values.

Now consider the problem of estimating the linear function

$$\phi = \mathbf{c}^T \mathbf{x}. \tag{2.6-30}$$

We want an unbiased estimator

$$\hat{\phi} = \mathbf{u}^T \hat{\mathbf{b}}, \tag{2.6-31}$$

so we require that

$$E(\mathbf{u}^T \hat{\mathbf{b}}) \equiv \mathbf{c}^T \mathbf{x} \qquad \text{or} \qquad \mathbf{u}^T \mathbf{A}\mathbf{x} \equiv \mathbf{c}^T \mathbf{x}$$

whatever the value of $\mathbf{x}$. Thus the vector $\mathbf{u}$ must satisfy

$$\mathbf{u}^T \mathbf{A} = \mathbf{c}^T$$

or

$$\mathbf{A}^T \mathbf{u} = \mathbf{c}. \tag{2.6-32}$$

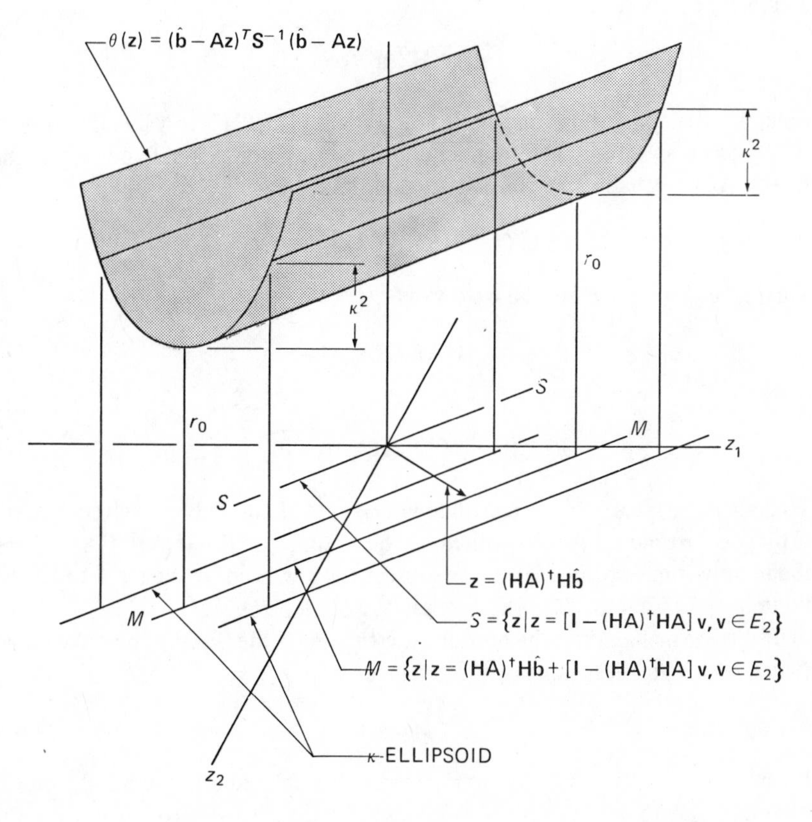

Figure 2.7. The residual surface and a κ-ellipsoid for a problem with $n = 2$ and $p = 1$.

This equation has solutions $\mathbf{u}$ if and only if the consistency condition (2.6-19) is satisfied; that is, if and only if

$$\mathbf{A}^T(\mathbf{A}^T)^\dagger \mathbf{c} = \mathbf{c}.$$

Now, from Eqs. (2.6-6) and (2.6-1) it follows that

$$\mathbf{A}^T(\mathbf{A}^T)^\dagger = \mathbf{A}^T(\mathbf{A}^\dagger)^T = (\mathbf{A}^\dagger \mathbf{A})^T = \mathbf{A}^\dagger \mathbf{A},$$

so a necessary and sufficient condition for the existence of a solution vector $\mathbf{u}$ is that

$$\mathbf{A}^\dagger \mathbf{A}\mathbf{c} = \mathbf{c}. \tag{2.6-33}$$

We now define a new variable $\mathbf{u}'$ by

$$\mathbf{u} = \mathbf{H}^T \mathbf{u}', \tag{2.6-34}$$

and rewrite (2.6-32) in the form

$$\mathbf{A}^T \mathbf{H}^T \mathbf{u}' = \mathbf{c}. \tag{2.6-35}$$

Since $\mathbf{H}^T$ is a nonsingular matrix, it follows that (2.6-32) is solvable if and only if (2.6-35) is solvable. Therefore, an unbiased estimator exists if and only if the matrix $\mathbf{A}^T\mathbf{H}^T$ satisfies the consistency condition

$$(\mathbf{A}^T \mathbf{H}^T)(\mathbf{A}^T \mathbf{H}^T)^\dagger \mathbf{c} = \mathbf{c}.$$

This last equation can also be written as

$$(\mathbf{HA})^\dagger (\mathbf{HA})\mathbf{c} = \mathbf{c}$$

or as

$$[\mathbf{I} - (\mathbf{HA})^\dagger (\mathbf{HA})]\mathbf{c} = \mathbf{0}. \tag{2.6-36}$$

Thus there exists an unbiased estimate of $\phi = \mathbf{c}^T\mathbf{x}$ if and only if the vector $\mathbf{c}$ is orthogonal to the subspace spanned by the columns of $[\mathbf{I} - (\mathbf{HA})^\dagger(\mathbf{HA})]$; that is, if and only if $\mathbf{c}$ is orthogonal to the subspace S. If $\mathbf{c}$ is not orthogonal to S, then no unbiased estimator exists and we say the ϕ is *inestimable.*

If ϕ is estimable, then the best linear estimator is the linear unbiased estimator that gives the smallest value of the variance

$$\sigma^2(\hat{\phi}) = \mathbf{u}^T \mathbf{S}\mathbf{u}. \tag{2.6-37}$$

It can be shown [10] that the minimal-variance unbiased estimator is the one corresponding to

$$\mathbf{u}^T = \mathbf{c}^T(\mathbf{HA})^\dagger \mathbf{H}. \tag{2.6-38}$$

The estimator itself can also be written

$$\hat{\phi} = \mathbf{c}^T \hat{\mathbf{x}}_{\mathbf{mv}} \tag{2.6-39}$$

where

$$\hat{\mathbf{x}}_{mv} = (\mathbf{HA})^{\dagger} \mathbf{H}\hat{\mathbf{b}} \tag{2.6-40}$$

is just the minimal-length solution of the normal equations. Interval estimates for ϕ are obtained in the same manner as for the case where **A** has full rank; that is, by means of the support planes of the κ-ellipsoid. Of course, if ϕ is inestimable the support planes orthogonal to **c** will intersect the κ-ellipsoid at infinity and the will be no finite confidence interval for ϕ.

References

1 R. L. Plackett, A historical note on the method of least squares, *Biometrika* **36** (1949), 458–460.
2 T. A. Magness and J. B. McGuire, Comparison of least squares and minimum variance estimates of regression parameters, *Ann. Math. Statist.* **33** (1962), 462–470.
3 Gene H. Golub, Comparison of the variance of minimum variance and weighted least squares regression coefficients, *Ann. Math. Statist.* **34** (1963), 984–991.
4 Robert V. Hogg and Allen T. Craig, *Introduction to Mathematical Statistics.* Macmillan, New York, 1965.
5 T. W. Anderson, *An Introduction to Multivariate Statistical Analysis.* Wiley, New York, 1958.
6 A. M. Mood and F. A. Graybill, *Introduction to the Theory of Statistics.* McGraw-Hill, New York, 1963.
7 Henry Scheffe, *The Analysis of Variance.* Wiley, New York, 1959.
8 Alston S. Householder, *The Theory of Matrices in Numerical Analysis,* Chapter 2. Random House (Blaisdell), New York, 1964.
9 R. Penrose, A generalized inverse for matrices, *Proc. Cambridge Phil. Soc.* **51** (1955), 406–413.
10 Charles M. Price, The matrix pseudoinverse and minimal variance estimates, *SIAM Rev.* **6** (1964), 115-120.

Chapter 3

A Thought Experiment

3.1 Introduction

This chapter presents an imaginary thought experiment that is typical of those requiring data analysis before the result can be presented in meaningful form. The purpose is to introduce in a heuristic manner the basic ideas that will be developed more formally in succeeding chapters. The material in this chapter may be skipped with no loss to our argument; in fact, the more abstract mathematician will find the imprecise word usage rather annoying.

3.2 The Experiment

Consider the experiment illustrated in Figure 3.1. An observer views a source of visible radiation from a candle through a filter. He uses an apparatus that quantitatively measures the intensity of the radiation that is passed through the filter.

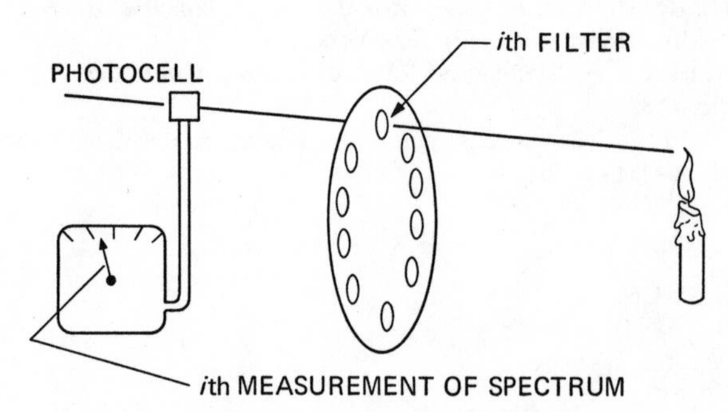

Figure 3.1. A gedanken experiment.

Let us suppose that he performs the observation for each of eleven filters, one after the other, and obtains "readings" having values $b_1, b_2, \ldots, b_{11}$, respectively.

We also suppose that the experimenter knows in advance the transmission of the filters as a function of the wavelength of light. The transmission functions for this gedanken experiment are shown in Figure 3.2.

The experimenter, whom we will hereafter call E, does not know anything in advance about the unknown spectrum, which we will denote by $x(s)$, except that it must be nonnegative at all wavelengths. He would like to use his observations $b_1, b_2, \ldots, b_{11}$ to infer as much information about the spectrum as he can, but

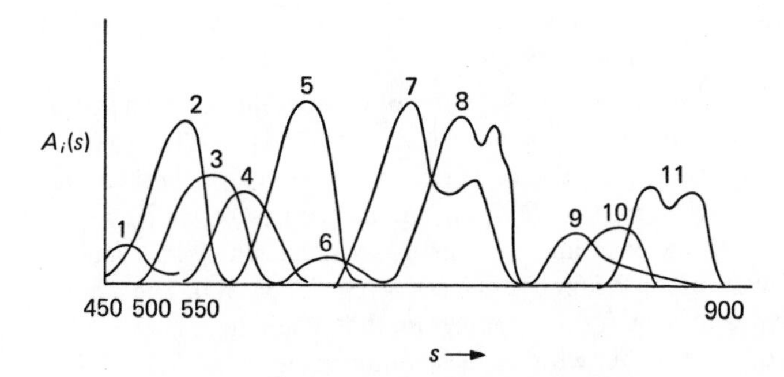

Figure 3.2. Filter transmission functions.

he quickly comes to the realization that with only eleven measurements he can never reconstruct the spectrum in all its detail. In fact, none of the filters transmit any light outside the range $s = 450$ to $s = 900$, so that his measurements tell him absolutely nothing about the spectrum from 0 to 450 or over 900.

3.3 The Wished-for Filters

Let us suppose that E is a conscientious experimenter and that the nature of his work is such that it is highly desirable to make valid predictions about the results. He, therefore, has a reluctance to make any predictions that cannot be substantiated by his measurements.

One possibility for E is to simply report his observation $b_1, b_2, \ldots, b_{11}$, together with the transmission functions $A_1(s), A_2(s), \ldots, A_{11}(s)$ for his eleven filters. Then, anybody receiving E's data would know as much about the spectrum of the candle as E himself.

But E is unhappy about doing this, since his filters have peculiar shapes and features. He feels intuitively that a certain amount of data reduction ought to be done to make the results more easily intelligible to others. E wishes that, although he cannot build a perfect instrument, he could build one of about the same caliber as the actual one but with more uniform filter transmission functions.

One way of looking at this problem is to assume that the experimenter imagines a "wish instrument." The wish instrument has characteristics that are similar to those of the actual instrument except that its response functions have more uniform characteristics. E realizes, however, that he cannot hope for a wish instrument that is too far removed from reality. He now wishes to transform numerically the actual data he has obtained on the real instrument so that they appear to have been taken on his wish instrument. In other words, he hopes to be able to estimate the result that he would have obtained on his wish instrument by intelligent manipulation of the results actually obtained. We have posed a problem of unfolding in the following form.

> An experimenter E makes a series of observations with an instrument whose response functions are $A_1(s), A_2(s), \ldots, A_{11}(s)$. The values of his observations, using each filter, are $b_1, b_2, \ldots, b_{11}$. We consider that the true unknown spectrum $x(s)$ is known only to nature, which we denote by N. However, the experimenter E imagines an imaginary instrument with a set of wish response functions $w_1(s), w_2(s), \ldots, w_{12}(s)$. The mathematical problem is to solve for the readings p_k that would have been made on the imaginary instrument with response functions $w_1(s), w_2(s), \ldots, w_{12}(s)$.

3.4 The Optimal Combination Estimation Method

There are a number of methods that E might use to determine the readings that his wish instrument would have given, although the true spectrum $x(s)$ is known only to N and cannot be determined from just eleven measurements. E could propose a polynomial model or some parametric formula. Based on his observations $b_1, b_2, \ldots, b_{11}$, he could then estimate the parameters of the model or the coefficients of the polynomial; but as we have observed, E is a very conscientious observer and wishes to avoid any assumptions that do not rest on a firm physical basis. Thus our experimenter is reluctant to use a polynomial or some other arbitrary model for $x(s)$.

One possibility that might occur to E is to select a particular set of wish response functions that resemble those of the actual instrument, such as the set shown in Figure 3.3. Rather than assume an unrealistic set of very narrow Gaussian functions, E might assume an imaginary set of functions that resemble those of the actual instrument, but a set whose functions are normalized and symmetric in shape. The actual instrument had widely varying normalization between response functions, and some response functions had multiple peaks as well. In addition to the symmetric peaked windows, E might also choose two other windows that are of special interest to him. Number twelve has constant height for all values of s. A measurement made with an instrument with a response function of this sort would be a measurement of the total integral of the spectrum over the entire region. The second special imaginary response function, number seven, is rectangular in shape and covers the interval from $s = 750$ to $s = 850$. A measurement made on an instrument with this response would respond only to the integral of the spectrum from $s = 750$ to $s = 850$.

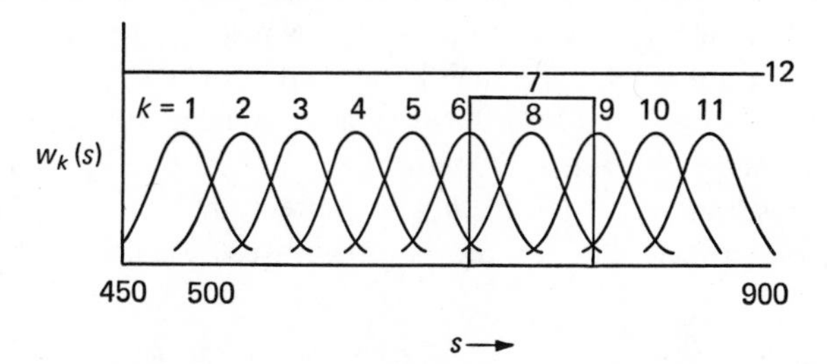

Figure 3.3. A set of realistic wish functions.

It has also occurred to E that it might be desirable to make all of his imaginary windows with a rectangular histogram-type shape, so that each imaginary window might cover a narrow range of s over the entire region of interest. By having only as many rectangular windows as there are observations, a completely determined set of equations would result, which would have a completely determined solution. The problem is that the integral equation, which relates the spectrum $x(s)$ to b, must be reduced to a finite form in order to obtain a small number of equations, and this process itself is subject to error.

There are, however, two conservative methods that E may use that do not depend on any arbitrary assumptions. E notices that he can quickly find an upper bound for the response of an instrument that has a rectangular response function. Figure 3.4 illustrates the method of finding an upper bound. The

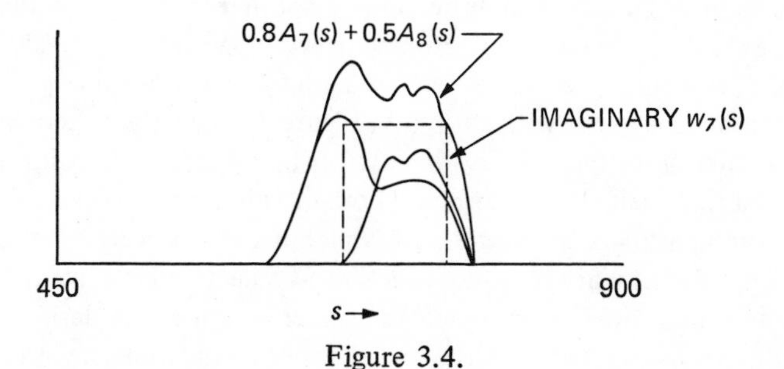

Figure 3.4.

rectangular imaginary response function is indicated in the figure by a dashed curve. Also illustrated are two of the actual response functions drawn on the same scale. By constructing the indicated sum of the two actual response functions with the weighting coefficients shown, the upper solid curve is obtained, which is everywhere greater than the desired imaginary response function. Since it is known that an instrument with the actual response function yields an observation of $b_1, b_2, \ldots, b_{11}$ when observing the spectrum $x(s)$, an instrument with the response shown by the upper curve would have yielded a result given by the sum of products of the corresponding real response functions and the appropriate constants. Thus, although E does not know the spectrum $x(s)$, he may still find an upper bound for the desired rectangular response function by observations made on his actual instrument.

If E's observations of $b_1, b_2, \ldots, b_{11}$ are correct, then the resulting upper bound is absolutely safe. Since E knows also with 100% certainty that the

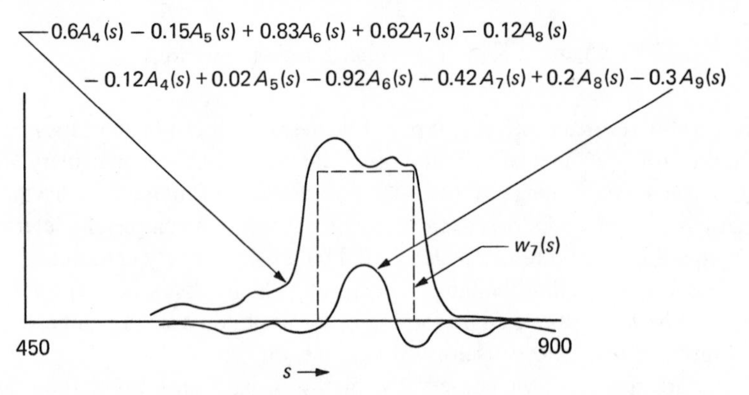

Figure 3.5. Optimal upper and lower "windows".

spectrum $x(s)$ must everywhere be nonnegative, he can also give a 100% safe lower bound of 0. Thus he establishes an interval that must contain the desired result p_7. In fact, E can come up with a narrower confidence interval. Let us suppose that E wishes to obtain as narrow a confidence interval as possible that will still contain the result with certainty. Figure 3.5 shows another combination of response functions that is everywhere larger than the same imaginary response function and that yields a smaller upper bound. Similarly, the figure shows a still different combination of actual response functions that is everywhere smaller than the desired imaginary response function. If an actual instrument had had this response, then it would have yielded a lower bound to the desired one. Just as in the case of the upper bound, the response of an instrument with any synthetic response function can be estimated from the actual $b_1, b_2, \ldots, b_{11}$ by multiplying the appropriate values of b by the coefficients of combination of the actual response functions that make up the synthetic response functions.

We will see in succeeding chapters that the mathematical formulation of E's best lower bound and best upper bound is a linear programming problem. Linear programming is a term that is used to describe a particular type of mathematical calculation that minimizes or maximizes a linear function of s under certain inequality constraints. The inequality constraints discussed here, in terms of unfolding, are the requirements that the combination response functions be either above the imaginary response functions $w(s)$ for the upper bound or below $w(s)$ for the lower bound.

3.5 Nonnegativity

In the method of optimal combinations for finding upper and lower bounds to the value that would have been obtained on the imaginary instrument, we did not make explicit use of the nonnegativity of the spectrum $x(s)$. However, the

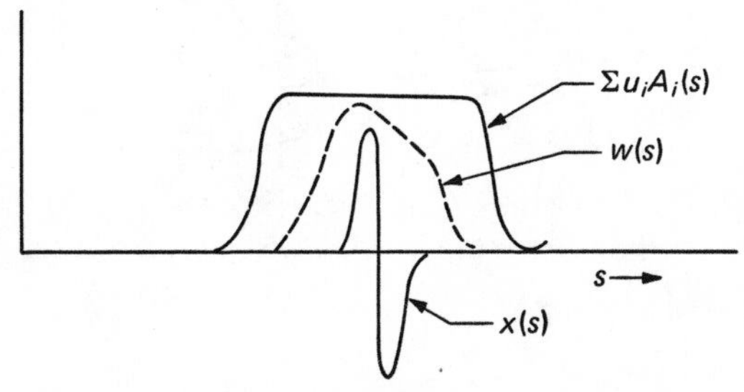

Figure 3.6. Mixed negative and positive spectra.

nonnegativity of $x(s)$ is necessary to assure the success of this method. Nonnegativity in this case is a physical necessity arising from the fact that the optical spectrum from the candle consists of packets of discrete quanta. There is no way to transport negative energy. It is easy to see that if nature had allowed negative components in spectra, as well as positive components, the situation would have been much more complex. Consider, for example, the spectrum shown in the Figure 3.6; $x(s)$ here consists of a positive peak followed by a negative peak. Two response functions of the type we have been talking about are also shown. The upper one is everywhere larger than the lower. However, if two instruments with these response functions were used to make an observation upon $x(s)$, then the one with the upper response function would yield a value of 0, since the negative and positive peaks in the spectrum would just cancel out; whereas the lower of the two response functions would yield a positive value, since its response to the positive peak is larger than the response to the negative peak. Thus, the inequality relationship that guarantees that the upper response function produces a larger value of observation holds only in the case of a nonnegative spectrum.

3.6 Worst Spectrum Estimation

There is a second way of estimating values that would have been produced by instruments having an imaginary response function. In Fig. 3.7, we have indicated schematically the situation in which we are trying to unfold an underdetermined system of equations. The graph on the left represents the experimental observation of $b_1, b_2, \ldots, b_n$. The graph on the right represents the acceptable solutions, which we denote in a discrete space $x_1, x_2, \ldots, x_n$. We assume that we know an observed measurement point **b**, which we denote by a point in the *b*-space. For illustration, we have shown a set of possible **x** values in *x*-space as a shaded region in the right-hand graph. Each point in this set corresponds to a particular spectrum,

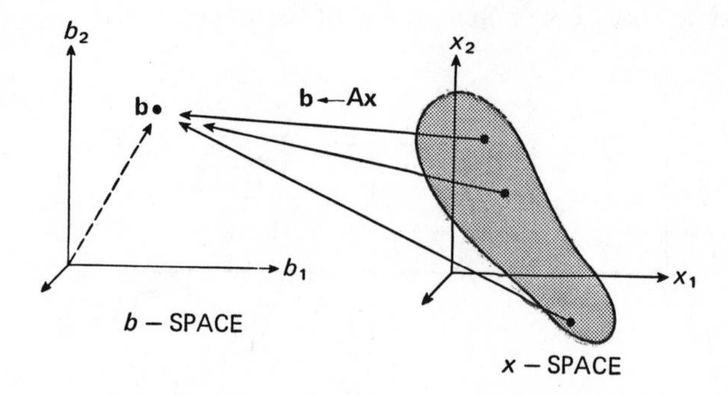

Figure 3.7. Acceptable solutions.

$x(s)$ in the continuous case. A spectrum may be considered experimentally acceptable if an observation made with the real instrument yields results consistent with the **b**-vector in *b*-space. Thus each point in the acceptable set of x solutions must transform to the single **b** point in *b*-space.

The relationship between the spectrum **x** and the observation **b** is shown symbolically on the graph by $\mathbf{b} \leftarrow \mathbf{Ax}$. The transformation from all points in the set of acceptable solutions yields a single value of **b**, but the inverse transformation from **b** to **x** is multivalued. This is typical of poorly conditioned or undetermined problems, and is responsible for the difficulty in unfolding. A conservative approach to unfolding is to take the entire set of solutions that are experimentally acceptable and to base interval estimates for the desired estimable function on the points in *x*-space that yield the largest and lowest value for the function. The particular approach to unfolding that we have taken involves an imaginary instrument with wish response functions $w_1(s), w_2(s), \ldots, w_k(s)$. Let us take a simple example of a w function for illustration in our discrete example. Suppose that $p = x_1 + 2x_2$. In other words, the window function in this case is a discrete vector with components (1, 2). In Fig. 3.8, values of constant p are shown by diagonal lines. These lines may be thought of as level planes in x-space on which values of p are constant. The inclination of these lines, of course, depend on the particular components of **w**. In the "worst spectrum" estimation method, in order to find an estimate for p, we find two points in the acceptable set. One point, denoted by $\mathbf{x}^{\max}$, yields the largest value of p, and another point, denoted by $\mathbf{x}^{\min}$, gives the smallest value of p. These two special spectra $\mathbf{x}^{\min}$ and $\mathbf{x}^{\max}$ are referred to as the worst spectra or the extreme spectra. Now, if it is known, a priori, that the spectrum **x** is nonnegative, these results can be immediately improved. In this case, $\mathbf{x}^{\min}$ and $\mathbf{x}^{\max}$ are constrained to positive values, and the largest and smallest value of p would be those that are shown in Fig. 3.8.

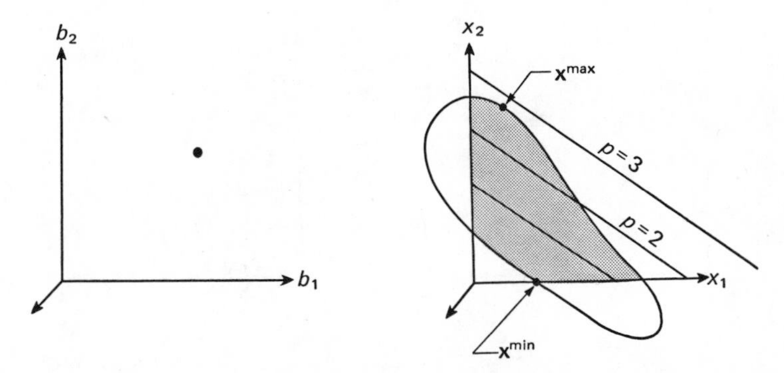

Figure 3.8. Extremal worst spectra for nonnegative constrained situation.

Both the method of optimal combinations (Section 3.4) and the method of worst spectra are examples of what is known as mini-max estimates. Mini-max estimates arise in a variety of contexts in military strategy, gambling, economics, and so forth. In our context, we may view unscrambling as a contest between nature N and the experimenter E. Only N "knows" the true spectrum $x(s)$. E must protect himself against the worst possible spectrum that N may contrive. E takes the position that he must minimize the maximum damage that N can do to his confidence intervals.

3.7 Errors in the Observations

Thus far, we have considered only the case where the observations $b_1, b_2, \ldots, b_{11}$ were measured without error. Usually, a certain error s_i is associated with each measured b_i. For simplicity, we assume that E can estimate the errors of his measurements of b_i such that $\hat{b}_i - s_i \leqslant \bar{b}_i \leqslant \hat{b}_i + s_i$ where $\hat{b}_i$ is measured by E, and $\bar{b}_i$ is the true value. Now the two methods that we previously discussed must be modified to take these errors into account. The worst spectrum approach can be modified as shown in Fig. 3.9. Here, the **b** measurement is shown not as a point, but as a box. The dimensions of the box represent the possible errors. It is known that the true value of **b** lies somewhere within the box. In this case, any solution **x** must be considered acceptable if it corresponds to any point within the **b** box. Mathematically, the problem of finding an upper bound and a lower bound to the value of p requires that two extreme spectra $\mathbf{x}^{\max}$ and $\mathbf{x}^{\min}$ be found within the allowed region, which result in the largest and smallest possible value for p.

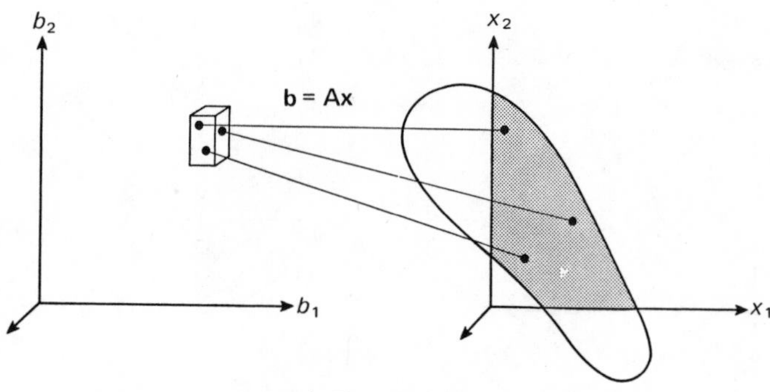

Figure 3.9.

Now let us consider what modifications must be made to the method of optimal combinations in order to account for the errors. Previously, without errors, the method of optimal combinations had the following formulation.

For an upper bound to the estimate, find a set of coefficients u_i such that $\Sigma u_i A_i(s)$ is an upper window and $\Sigma u_i b_i$ is minimized.

For the lower bound to the estimate, find the set of coefficients u_i such that $\Sigma u_i A_i(s)$ is a lower window and $\Sigma u_i b_i$ is maximized.

In order to account for the errors, we no longer wish to minimize or maximize $\Sigma u_i b_i$, but instead we wish to include the possible errors in the minimization of maximization procedures. We, therefore, reformulate the problem as follows.

For the upper bound to the estimate, find the set of coefficients u_i such that $\Sigma u_i A_i(s)$ is an upper window and $\Sigma u_i b_i + |u_i| s_i$ is minimized.

{For the lower bound to the estimate, find a set of coefficients u_i such that $\Sigma u_i A_i(s)$ is a lower window and $\Sigma u_i b_i - \Sigma |u_i| s_i$ is maximized.

The new term $\Sigma |u_i| s_i$ increases or decreases the estimate by the maximum error that could be made. Notice that the absolute value is necessary, since the errors do not necessarily cancel when one of the coefficients of combination u_i is negative.

In illustration, we will take the same case that we worked out for no errors, and determine the best upper and lower windows for measurements that have a 1% error. Compare Fig. 3.10 with Fig. 3.5 for no errors. Notice that the windows

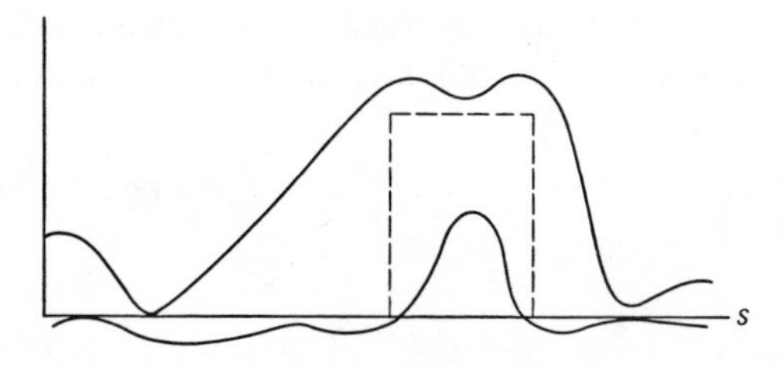

Figure 3.10. Upper and lower windows optimal for 1% errors.

have substantially different shapes. Instead of trying to find the fits that make the interval

$$[\Sigma u_i^{\mathrm{lo}} b_i \text{ to } \Sigma u_i^{\mathrm{up}} b_i]$$

as narrow as possible, consideration is also given to keeping the resulting errors small. Usually, this involves finding upper and lower functions that are further from the desired wish function.

Although the new windows are further from the desired windows, they yield an overall interval that is narrower. It is because these windows are optimal for a certain error in the observations **b**. This situation is made clearer by Table 3.1.

TABLE 3.1

	Confidence interval using	
Windows optimal for	$\mathbf{b}^{\mathrm{true}}$ and $s_i = 0$	$\mathbf{b}^{\mathrm{true}}$ and $s_i = 0.01 b_i$
No errors	(13.2, 16.2)	(9.2, 20.2)
1% errors	(12.2, 17.2)	(11.2, 18.2)

3.8 Random Errors

We have discussed cases with no errors in the observation and with errors in the observation, such that the true value of **b** is known to be within a box centered about the measured value of **b**. Neither of these situations is applicable to most measurements. Instead, a random error s_i is made for each point. Instead of obtaining a box that is known to contain the true value of **b**, a confidence region is obtained. This confidence region is defined so that if the experimental observations are repeated many times, for the same spectrum $x(s)$, the confidence region will include the true value of **b** on a certain fraction of the trials (typically 67% or 90%). If the errors in the observation are normally distributed, then the confidence regions in b-space are ellipsoids as shown in Fig. 3.11. This figure

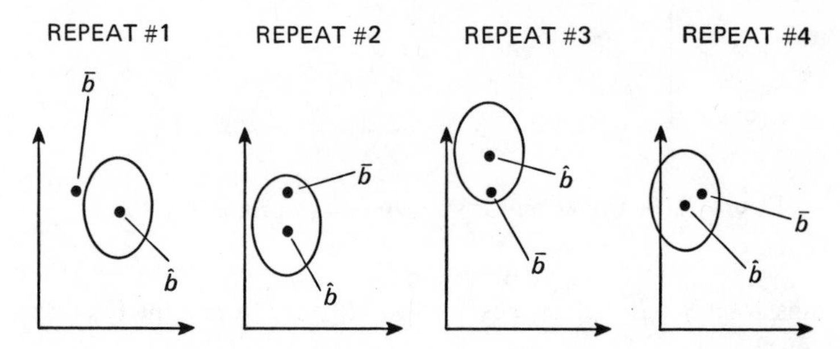

Figure 3.11. Statistical confidence region for measurement b.

shows a 67% confidence interval for several different repeated experiments. Approximately 2/3 of the confidence intervals include the true b. Note that there is no experimental way for the experimenter to verify that the confidence intervals he sets up actually cover the true **b** the correct fraction of the time, since the true **b** depends on the true **x**, which is known only to N.

CHAPTER 4

Constrained Linear Estimation

4.1 Introduction

In Section 2.6 we saw that the $m \times n$ linear system

$$\mathbf{K}\bar{\mathbf{x}} = \bar{\mathbf{b}} \tag{4.1-1}$$

is solvable if and only if $\mathbf{K}$ and $\bar{\mathbf{b}}$ satisfy the consistency condition

$$\mathbf{K}\mathbf{K}^{\dagger}\bar{\mathbf{b}} = \bar{\mathbf{b}}$$

where $\mathbf{K}^{\dagger}$ is the generalized inverse of $\mathbf{K}$. In this chapter, we assume that $\mathbf{K}$ and $\bar{\mathbf{b}}$ do satisfy the consistency condition and consider problems of the form

$$\|\mathbf{K}\hat{\mathbf{x}} - \hat{\mathbf{b}}\|_2 = \min_{\mathbf{x}} \|\mathbf{K}\mathbf{x} - \hat{\mathbf{b}}\|_2 \tag{4.1-2}$$

where $\hat{\mathbf{b}}$ denotes a sample from a distribution of possible $\mathbf{b}$-vectors. This $\mathbf{b}$ distribution is associated with the vector $\bar{\mathbf{b}}$, which is assumed to exist but is unknown to us. Equation (4.1-1) can also be written

$$\mathbf{K}\bar{\mathbf{x}} = \hat{\mathbf{b}} - \hat{\mathbf{e}} \tag{4.1-3}$$

where $\hat{\mathbf{e}}$ is the vector defined by

$$\hat{\mathbf{e}} = \hat{\mathbf{b}} - \bar{\mathbf{b}}. \tag{4.1-4}$$

The vector $\hat{\mathbf{e}}$ can be regarded as an uncertainty or error in the vector $\hat{\mathbf{b}}$. Thus, each sample $\hat{\mathbf{b}}$ from the $\mathbf{b}$ distribution corresponds to a sample $\hat{\mathbf{e}}$ from an error distribution. We will consider the following three types of error distributions.

$$\text{I.} \quad \sum_{i=1}^{m} \frac{|e_i|}{s_i} \leqslant 1, \tag{4.1-5}$$

$$\text{II.} \quad \sum_{i=1}^{m} \frac{|e_i|^2}{s_i^2} \leqslant 1, \tag{4.1-6}$$

$$\text{III.} \quad \max_{1 \leqslant i \leqslant m} \left\{ \frac{|e_i|}{s_i} \right\} \leqslant 1, \tag{4.1-7}$$

where the s_i are given constants. These three regions are illustrated graphically for the case $m = 2$ in Fig. 4.1. In each case, the error vector is arbitrarily distributed in the region defined by the inequality constraint (the shaded area in the figure).

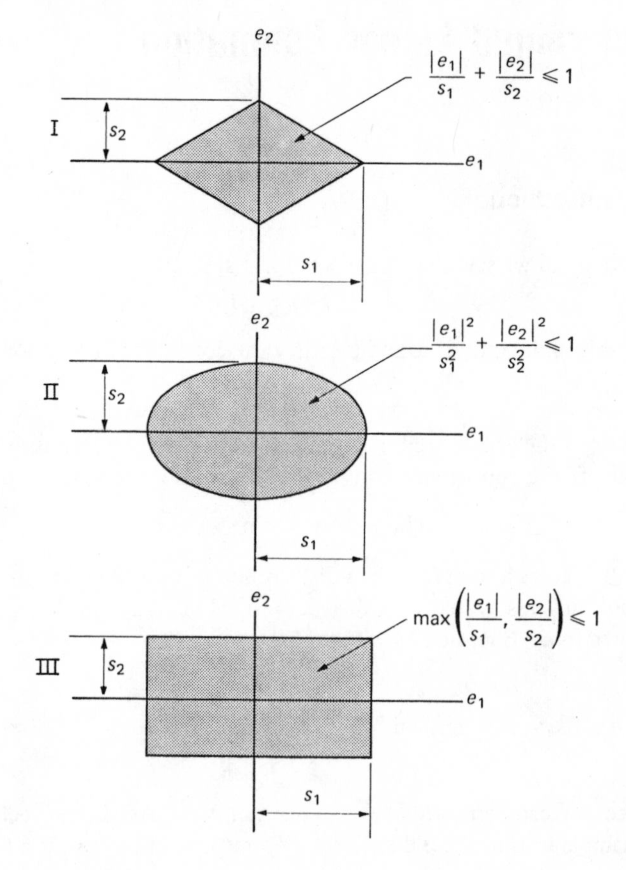

Figure 4.1. Three kinds of error distributions.

Each of the foregoing error distributions induces in x-space a corresponding distribution of solution vectors $\hat{\mathbf{x}}$. In the general case, the columns of $\mathbf{K}$ will not be linearly independent, so there will be many solution vectors $\hat{\mathbf{x}}$ corresponding to any given sample $\hat{\mathbf{b}}$. For a given $\hat{\mathbf{b}}$, the quantity to be minimized is

$$\|\mathbf{Kx} - \hat{\mathbf{b}}\|_2 = +[(\mathbf{Kx} - \hat{\mathbf{b}})^T (\mathbf{Kx} - \hat{\mathbf{b}})]^{1/2}, \tag{4.1-8}$$

from which it follows that

$$\|\mathbf{K}\hat{\mathbf{x}} - \hat{\mathbf{b}}\|_2 = +\,[\min_{\mathbf{x}} (\hat{\mathbf{b}} - \mathbf{K}\mathbf{x})^T (\hat{\mathbf{b}} - \mathbf{K}\mathbf{x})]^{1/2}. \tag{4.1-9}$$

In Section 2.6, we saw that the quadratic form

$$\theta(\mathbf{z}) = (\hat{\mathbf{b}} - \mathbf{A}\mathbf{z})^T \mathbf{S}^{-1} (\hat{\mathbf{b}} - \mathbf{A}\mathbf{z}),$$

where $\mathbf{S}^{-1}$ is a positive-definite matrix with a factorization of the form

$$\mathbf{S}^{-1} = \mathbf{H}^T \mathbf{H} = \mathbf{H}^2,$$

is minimized by any vector $\hat{\mathbf{z}}$ of the form

$$\hat{\mathbf{z}} = (\mathbf{H}\mathbf{A})^\dagger \mathbf{H}\hat{\mathbf{b}} + [\mathbf{I} - (\mathbf{H}\mathbf{A})^\dagger \mathbf{H}\mathbf{A}]\,\mathbf{v}$$

where $\mathbf{v}$ is an arbitrary vector in the same space as the rows of $\mathbf{A}$. Thus, for a given sample $\hat{\mathbf{b}}$, the expression in (4.1-8) is minimized by any vector $\hat{\mathbf{x}}$ of the form

$$\hat{\mathbf{x}} = \mathbf{K}^\dagger \hat{\mathbf{b}} + [\mathbf{I} - \mathbf{K}^\dagger \mathbf{K}]\,\mathbf{v} \tag{4.1-10}$$

where $\mathbf{v}$ is an arbitrary n-vector. In the special case where $\mathbf{K}$ has linearly independent columns, there is only one solution

$$\hat{\mathbf{x}} = \mathbf{K}^\dagger \hat{\mathbf{b}} = (\mathbf{K}^T \mathbf{K})^{-1} \mathbf{K}^T \hat{\mathbf{b}},$$

and this solution satisfies the system of equations

$$\mathbf{K}\hat{\mathbf{x}} = \hat{\mathbf{b}} \tag{4.1-11}$$

exactly, so that the square of the minimum value of (4.1-8) is

$$r_0 = \|\mathbf{K}\hat{\mathbf{x}} - \hat{\mathbf{b}}\|^2 = 0. \tag{4.1-12}$$

If the columns of $\mathbf{K}$ are not linearly independent. then the system (4.1-11) may or may not be a consistent system. If the vector $\hat{\mathbf{b}}$ is such that the consistency condition is satisfied, then all of the solutions (4.1-10) satisfy (4.1-11) exactly, and the minimum value r_0 is equal to 0. If the consistency condition is not satisfied, then Eq. (4.1-11) does not have any exact solutions, and the solutions $\hat{\mathbf{x}}$ to (4.1-2) produce a minimum value of

$$r_0 = \|\mathbf{K}\hat{\mathbf{x}} - \hat{\mathbf{b}}\|^2 = \hat{\mathbf{b}}^T [\mathbf{I} - \mathbf{K}\mathbf{K}^\dagger]\,\hat{\mathbf{b}}. \tag{4.1-13}$$

In either case, the set of all solutions

$$\{\hat{\mathbf{x}}\} = \{\mathbf{K}^\dagger \hat{\mathbf{b}} + [\mathbf{I} - \mathbf{K}^\dagger \mathbf{K}]\,\mathbf{v} \,|\, \mathbf{v} \text{ arbitrary}\} \tag{4.1-14}$$

is a coset of the subspace S of all vectors of the form $[\mathbf{I} - \mathbf{K}^\dagger \mathbf{K}]\,\mathbf{v}$.

We will not in general be primarily interested in the solution vectors themselves, but rather we seek to estimate the quantity

$$\bar{\mathbf{p}} = \mathbf{W}^T \bar{\mathbf{x}} = \begin{pmatrix} \mathbf{w}_1^T \\ \mathbf{w}_2^T \\ \cdot \\ \cdot \\ \cdot \\ \mathbf{w}_k^T \end{pmatrix} \bar{\mathbf{x}} \tag{4.1-15}$$

where $\mathbf{W}^T$ is a given $k \times n$ window matrix whose rows are the n-vectors $\mathbf{w}_1^T, \mathbf{w}_2^T, \ldots, \mathbf{w}_k^T$. (Note that if it is $\bar{\mathbf{x}}$ we are interested in, then we can choose $\mathbf{W}^T$ to be

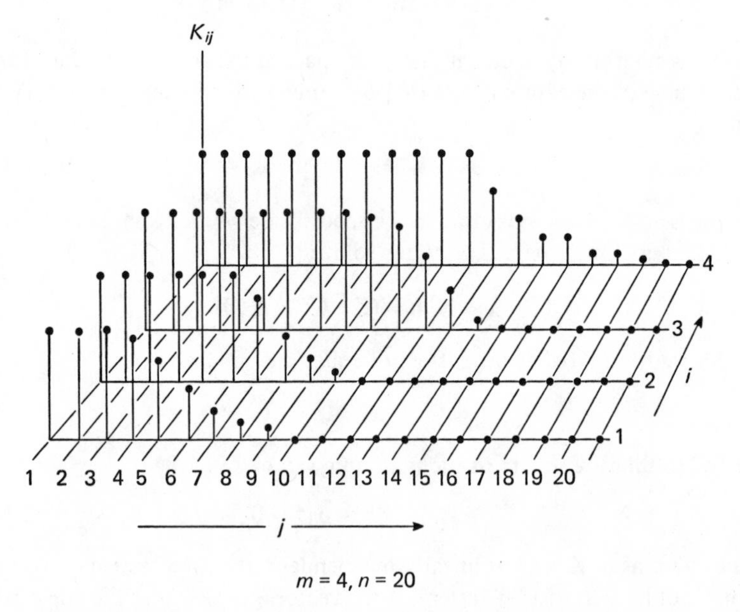

Figure 4.2. The system matrix for a 4 x 20 (underdetermined) linear system.

the $n \times n$ identity matrix.) If the columns of **K** are not linearly independent, then there will be many widely differing estimates

$$\hat{\mathbf{p}} = \mathbf{W}^T \hat{\mathbf{x}} \tag{4.1-16}$$

for $\bar{\mathbf{p}}$ corresponding to a given sample vector $\hat{\mathbf{b}}$. Even when the columns of **K** are linearly independent, if the system (4.1-11) is poorly conditioned, then small changes in $\hat{\mathbf{b}}$ can produce very large changes in $\hat{\mathbf{x}}$ and hence very large changes in $\hat{\mathbf{p}}$. In either case, the $\hat{\mathbf{b}}$ distribution defined by Eq. (4.1-4) and any one of the error distributions I, II, III will induce a widely varying $\hat{\mathbf{x}}$ distribution and hence a widely varying distribution of the estimates $\hat{\mathbf{p}}$. It will be our purpose in this chapter to consider how a priori knowledge about the solution vector $\bar{\mathbf{x}}$ can be used to decrease the amount of variation in the estimates $\hat{\mathbf{p}}$ that is induced by the variations $\hat{\mathbf{e}}$ in $\hat{\mathbf{b}}$. The a priori knowledge that we will use is simply that all

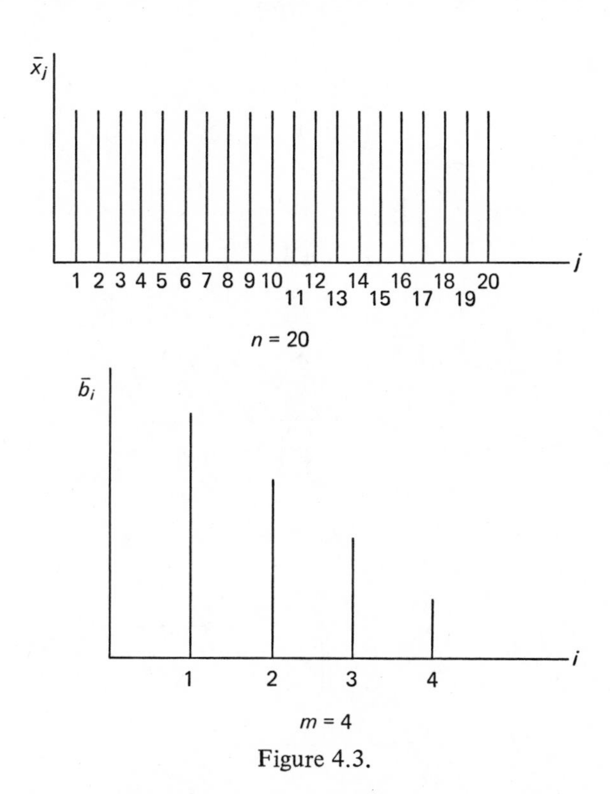

Figure 4.3.

of the elements in the solution vector must be nonnegative. That is, no estimate $\hat{\mathbf{p}}$ is acceptable unless it corresponds to a solution vector $\hat{\mathbf{x}}$ whose elements satisfy

$$\hat{x}_i \geqslant 0, \qquad i = 1, 2, \ldots, n. \tag{4.1-17}$$

An example of the kind of problem we are dealing with is illustrated in Figs. 4.2, 4.3, and 4.4. In Fig. 4.2, the elements K_{ij} of a 4×20 matrix $\mathbf{K}$ are plotted as heights above the i,j-plane, and Fig. 4.3 is a plot of the elements $\bar{b}_i$ of the vector $\bar{\mathbf{b}}$, together with a plot of the elements of *one possible* solution vector $\bar{\mathbf{x}}$. Since the system is underdetermined, there will be many other solution vectors $\bar{\mathbf{x}}$, but we are only interested in those vectors for which all the $\bar{x}_j$ are nonnegative.

Figure 4.4 shows a plot of the elements of a 2 x 20 window matrix $\mathbf{W}^T$, together with a plot of the vector $\bar{\mathbf{p}}$ that is obtained by combining this matrix with the $\bar{\mathbf{x}}$-vector in Fig. 4.3. Again, since the system is underdetermined, there will be many possible such vectors $\bar{\mathbf{p}}$ corresponding to the given $\mathbf{W}^T$; but again the a priori information rules out all such vectors except those that can be obtained from a vector $\bar{\mathbf{x}}$, in which all elements are nonnegative.

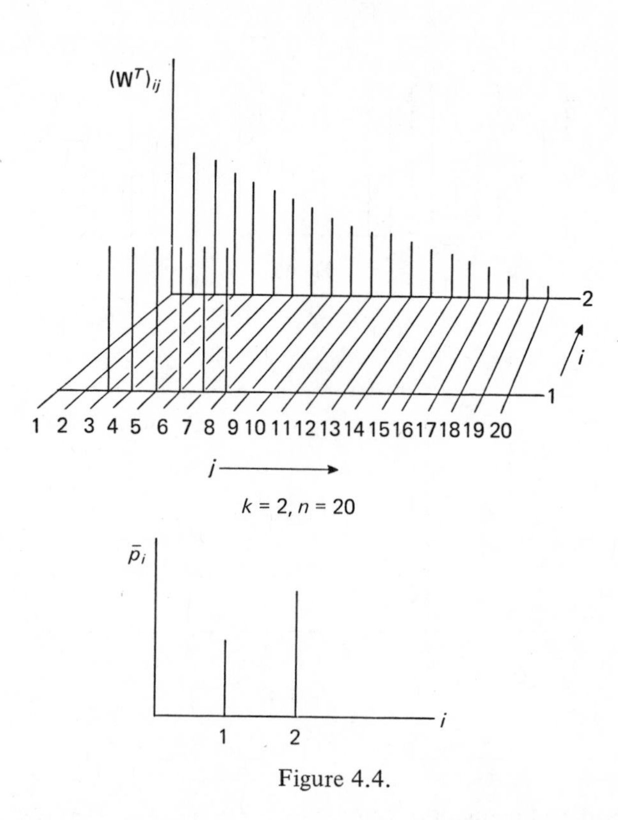

Figure 4.4.

Actually, the problem we are interested in is even more complicated than the one outlined in the previous paragraph because we do not actually know $\bar{\mathbf{b}}$. What we actually know is a sample vector $\hat{\mathbf{b}}$, or perhaps several such sample vectors, that were drawn from a population of $\mathbf{b}$-vectors that is arbitrarily distributed around the unknown $\bar{\mathbf{b}}$ in the region defined by one of the error distributions I, II, or III; that is, Eq. (4.1-2)–(4.1-4). The situation is illustrated in Fig. 4.5 for a problem where the error vector is distributed according to II. Only the first two components of the $\mathbf{b}$-vectors are plotted, but a similar picture would result from plotting any other pair of components. Figure 4.6 shows a plot of the first two components of the solution vectors **x** corresponding to the **b**-vectors in Fig. 4.5. The situation that is illustrated is one in which the system is underdetermined, so that each point in the b_1, b_2-space defines the coset of a linear subspace in the

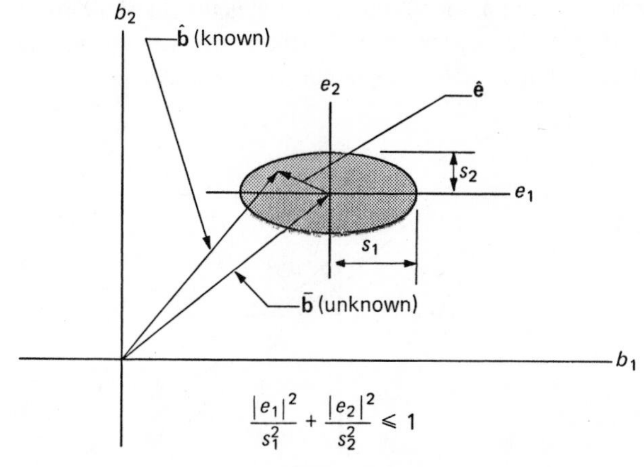

Figure 4.5.

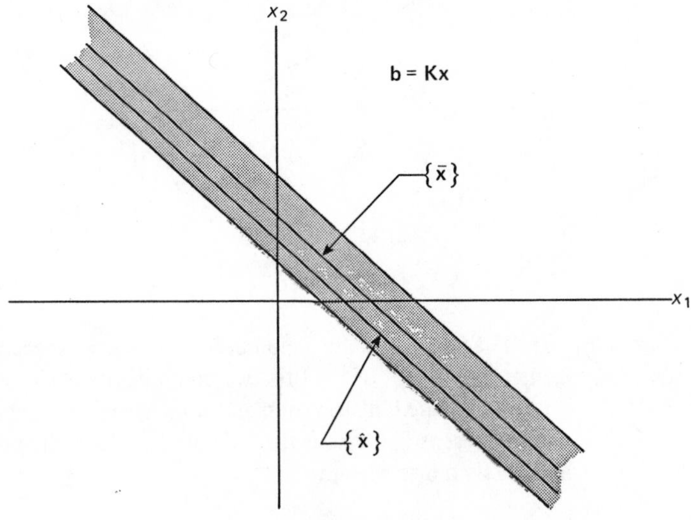

Figure 4.6.

x_1, x_2-space. The cosets $\{\bar{\mathbf{x}}\}$ and $\{\hat{\mathbf{x}}\}$ corresponding to the vectors $\bar{\mathbf{b}}$ and $\hat{\mathbf{b}}$ are shown. The shaded area is the set of all possible solution vectors that correspond to some **b**-vector in the shaded area defining the **b** distribution. Since $\bar{\mathbf{b}}$ is not known precisely, the subspace $\{\bar{\mathbf{x}}\}$ cannot be found; but for any sample vector $\hat{\mathbf{b}}$, the set $\{\hat{\mathbf{x}}\}$ could, in principle at least, be determined by using the generalized inverse and the expression

$$\{\hat{\mathbf{x}}\} = \{[\mathbf{K}^\dagger \hat{\mathbf{b}} + (\mathbf{I} - \mathbf{K}^\dagger \mathbf{K})\,\mathbf{v}] \text{ for } \mathbf{v} \text{ arbitrary}\}.$$

Of course, what we are really interested in is the set of all possible **p**-vectors defined by the **b**-vector distribution. Furthermore, we want to take into account the a priori information about the nonnegativity constraint on the elements of

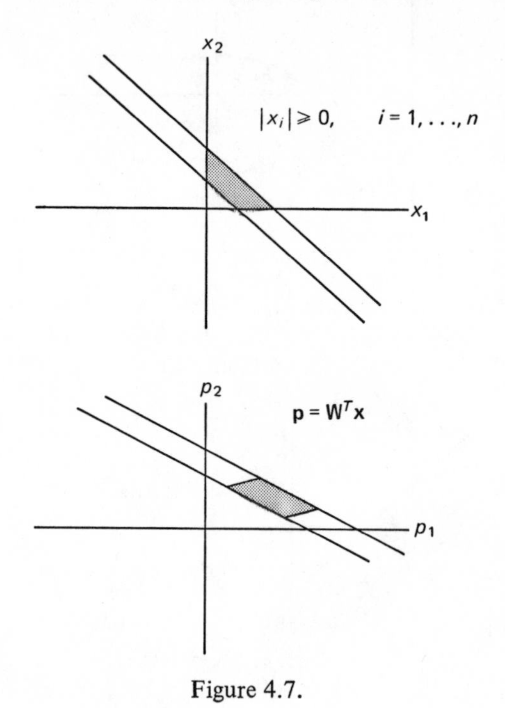

Figure 4.7.

the **x**-vectors. Figure 4.7 again shows the region defined in x_1, x_2-space by the **b** distribution, together with a plot of the first two components of the set of all **p**-vectors that are defined by the **b** distribution and some given window matrix $\mathbf{W}^T$. The shaded areas represent the regions that are allowed when the non-negativity constraint is taken into account.

4.2 The Constrained Estimation Technique

Each element of the vector $\hat{\mathbf{p}}$ is defined by

$$\hat{p}_i = \mathbf{w}_i^T \hat{\mathbf{x}}, \tag{4.2-1}$$

where $\mathbf{w}_i^T$ is the ith row of the window matrix $\mathbf{W}^T$. Ideally, we would like to know $\bar{p}_i$; but since we do not know $\bar{\mathbf{x}}$, we seek instead an interval

$$I_{\hat{p}_i} = [\hat{p}_i^{\,\mathrm{lo}}, \hat{p}_i^{\,\mathrm{up}}] \tag{4.2-2}$$

that is guaranteed to contain the value $\bar{p}_i$. Furthermore, we seek the narrowest possible interval that is consistent with the distribution of **b**-vectors and with the a priori nonnegativity constraint.

For a constant value of the quantity p_i, the expression

$$p_i = \mathbf{w}_i^T \mathbf{x} \tag{4.2-3}$$

defines a displaced linear subspace of x-space. We will refer to the rows of $\mathbf{W}^T$ as *window vectors,* and for any window vector $\mathbf{w}_i^T$, the surface defined by a constant value of p_i will be called a level surface for $\mathbf{w}_i^T$. Figure 4.8 shows

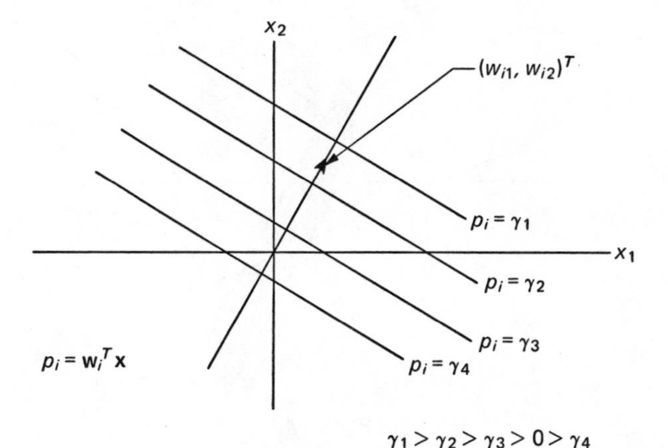

Figure 4.8.

the projection of a window vector and four corresponding level surfaces into the x_1, x_2-subspace of the x-space. A similar picture would be obtained if the projections were made into the subspace of any other pair of **x** components. The level surfaces are hyperplanes that are orthogonal to the window vector, and p_i has a constant value γ at every point on a given level surface. For the level surface that passes through the origin, the associated value of γ is 0, and the value of γ increases as the distance of the level surface from the origin increases along the direction of the vector $\mathbf{w}_i$. Similarly, the value of γ decreases from 0 as the distance of the level surface from the origin increases along the direction opposite from the direction of $\mathbf{w}_i$.

In order to find the interval $I_{\hat{p}_i}$, it is necessary to combine the idea of level surfaces with the idea of the induced solution set associated with the distribution of **b**-vectors. This is illustrated in two dimensions in Fig. 4.9. The shaded area is the set of possible solutions corresponding to the **b**-vector distribution. Any point in the solution set will lie on one of the level surfaces and the value γ associated with that level surface is the value of p_i associated with the point. It is clear that, in the situation illustrated, there will be no maximum or minimum value of p_i in the solution set. Thus, the interval $I_{\hat{p}_i}$ will be the trivial interval

$$I_{\hat{p}_i} = [-\infty, \infty].$$

This will be the usual situation in dealing with underdetermined systems, but it should be noted that even with underdetermined systems there are some window

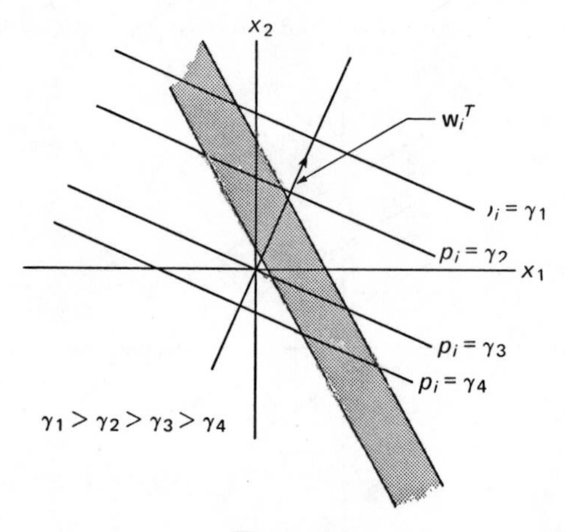

Figure 4.9.

vectors $\mathbf{w}_i^T$ that lead to a nontrivial interval $I_{\hat{p}_i}$. Such a situation is illustrated in Fig. 4.10. We saw in Section 2.6 that the requirement for a nontrivial interval in a situation like this is that $\mathbf{w}_i^T$ lie in the subspace spanned by those eigenvectors of $\mathbf{K}^T\mathbf{K}$ that correspond to the nonzero singular values of $\mathbf{K}$. If $\mathbf{w}_i^T$ has a nonzero component in the subspace spanned by the eigenvectors corresponding to a zero singular value, then $\mathbf{w}_i^T$ will have components parallel to the infinite dimensions of the solution set, and thus we can make p_i arbitrarily large by choosing some vector $\mathbf{x}$ in the set. This is the situation that is shown in Fig. 4.9.

We have just seen that most window vectors give functions that are inestimable for an underdetermined system; that is, they give a trivial interval estimate $I_{\hat{p}_i}$ for the corresponding function $\bar{p}_i$. But when the a priori nonnegativity constraint is taken into account, this regrettable situation is often remedied. This occurs when the intersection of the **x**-vector solution set and the positive orthant is a body of finite size in **x**-space. The situation is illustrated for x_1, x_2-space in Fig. 4.11. The shaded region is the constrained set of possible solution vectors and the minimum and maximum values that p_i can assume in this set are just the values

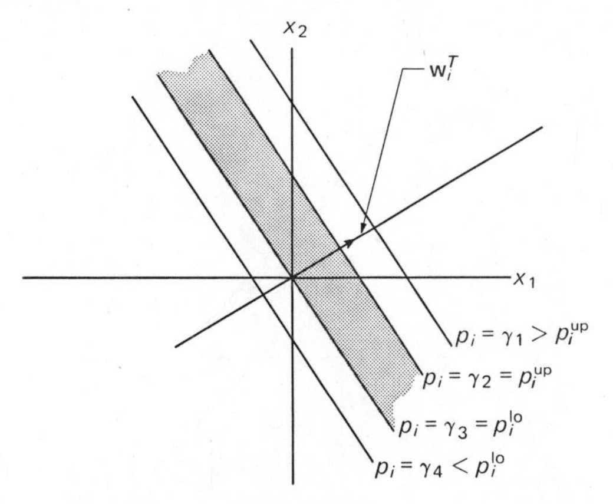

Figure 4.10.

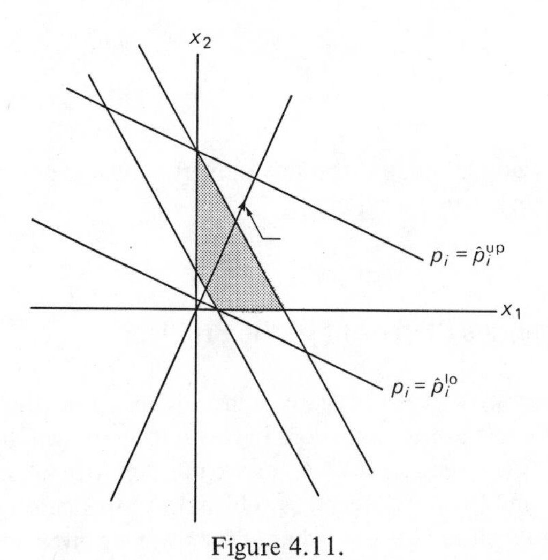

Figure 4.11.

$\hat{p}_i^{\text{lo}}$ and $\hat{p}_i^{\text{up}}$ that are associated with the level surfaces through the extreme points of the constraint region relative to the window vector. These level surfaces are just the planes of support of the constraint region in the direction of the window vector. The interval

$$I\hat{p}_i = [\hat{p}_i^{\text{lo}}, \hat{p}_i^{\text{up}}]$$

is guaranteed to contain the value $\bar{p}_i$. Figure 4.12 shows the region in p_1, p_2-space that is defined by the constrained solution vector set and the first two rows of the window matrix $\mathbf{W}^T$. The end points of the interval $I_{\hat{p}_1}$ are the extreme

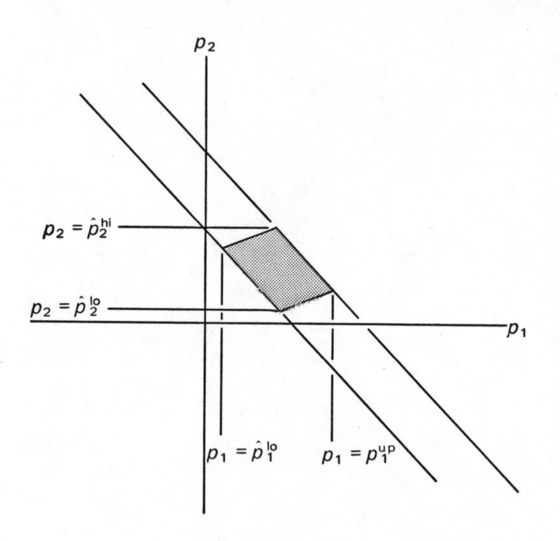

Figure 4.12.

points of this region in the p_1 direction. Similarly, the end points of $I_{\hat{p}2}$ are the extreme points in the p_2 direction.

4.3 An Algebraic Statement of the Problem

In the two preceding sections, we gave an intuitive geometrical formulation of the problem of constrained estimation. In this section, we will formulate the problem more exactly and show that, for each of the three kinds of error distributions I, II, and III, it reduces to a problem in mathematical programming.

The basic problem can be restated as follows. We are given a vector $\hat{\mathbf{b}}$, which is a sample drawn from an arbitrary distribution of $\mathbf{b}$-vectors satisfying

$$\|\mathbf{b} - \bar{\mathbf{b}}\|_{h,\mathbf{s}} = \|\mathbf{e}\|_{h,\mathbf{s}} \leqslant 1 \tag{4.3-1}$$

where

$$\mathbf{s} = \left(\frac{1}{s_1}, \frac{1}{s_2}, \ldots, \frac{1}{s_m}\right)^T$$

is a given constant vector, h is one of the three numbers $1, 2, \infty$, and the norm $\|\mathbf{e}\|_{h,\mathbf{s}}$ is defined by

$$\|\mathbf{e}\|_{h,\mathbf{s}} = \left[\sum_{i=1}^{m} \frac{|e_i|^h}{s_i^h}\right]^{1/h}. \qquad (4.3\text{-}2)$$

That is, $\|\mathbf{e}\|_{h,\mathbf{s}}$ is just a generalized order-h Hölder norm of the error vector $\mathbf{e}$. In general, we could let h have any integer value between 1 and ∞; but the most often-used values for h are 1, 2, and ∞, and these three values give just the three kinds of error vector distributions we are interested in; that is,

$$\text{I.} \quad h=1: \quad \|\mathbf{e}\|_{1,\mathbf{s}} = \sum_{i=1}^{m} \frac{|e_i|}{s_i}; \qquad (4.3\text{-}3)$$

$$\text{II.} \quad h=2: \quad \|\mathbf{e}\|_{2,\mathbf{s}} = \sum_{i=1}^{m} \frac{|e_i|^2}{s_i^2}; \qquad (4.3\text{-}4)$$

$$\text{III.} \quad h=\infty: \quad \|\mathbf{e}\|_{\infty,\mathbf{s}} = \max_{1 \leqslant i \leqslant m} \left\{\frac{|e_i|}{s_i}\right\}. \qquad (4.3\text{-}5)$$

Using set notation, we can say that $\hat{\mathbf{b}}$ is a vector picked at random from the set

$$B_{h,\mathbf{s}}(\bar{\mathbf{b}}) = \{\mathbf{b} \mid \|\mathbf{b} - \bar{\mathbf{b}}\|_{h,\mathbf{s}} \leqslant 1\}. \qquad (4.3\text{-}6)$$

We are also given an $m \times n$ matrix $\mathbf{K}$ that in general will not have linearly independent columns. If

$$\text{rank}\,(\mathbf{K}) = r,$$

then the set

$$\langle \mathbf{K} \rangle = \{\mathbf{b} \mid \mathbf{b} = \mathbf{K}\mathbf{x},\ \mathbf{x} \text{ arbitrary}\}$$

is an r-dimensional subspace of m-space. It is just the subspace spanned by the columns of the matrix $\mathbf{K}$. For any vector $\mathbf{b}$ in $\langle \mathbf{K} \rangle$, the consistency condition

$$\mathbf{K}\mathbf{K}^{\dagger}\mathbf{b} = \mathbf{b} \qquad (4.3\text{-}7)$$

is satisfied and the system

$$\mathbf{K}\mathbf{x} = \mathbf{b} \qquad (4.3\text{-}8)$$

is exactly solvable. For each such $\mathbf{b}$, the corresponding set of solutions is

$$\{\mathbf{x}(\mathbf{b})\} = \{\mathbf{x} = \mathbf{K}^{\dagger}\mathbf{b} + (\mathbf{I} - \mathbf{K}^{\dagger}\mathbf{K})\mathbf{v} \mid \mathbf{v} \text{ arbitrary}\}. \qquad (4.3\text{-}9)$$

We assume that the vector $\bar{\mathbf{b}}$ lies in the subspace $\langle \mathbf{K} \rangle$, so that the system

$$\mathbf{K}\bar{\mathbf{x}} = \bar{\mathbf{b}} \qquad (4.3\text{-}10)$$

is exactly solvable, but the sample vector $\hat{\mathbf{b}}$ will, in general, not lie in $\langle \mathbf{K} \rangle$, so that the system

$$\mathbf{Kx} = \hat{\mathbf{b}}$$

is not exactly solvable. The least squares minimization problem

$$\|\mathbf{K}\hat{\mathbf{x}} - \hat{\mathbf{b}}\|_2 = \min_{\mathbf{x}} \|\mathbf{Kx} - \hat{\mathbf{b}}\|_2 \tag{4.3-11}$$

is, however, solvable, and the solution set is

$$\{\hat{\mathbf{x}}(\hat{\mathbf{b}})\} = \{\hat{\mathbf{x}} = \mathbf{K}^\dagger \hat{\mathbf{b}} + (\mathbf{I} - \mathbf{K}^\dagger \mathbf{K})\,\mathbf{v} \mid \mathbf{v} \text{ arbitrary}\}. \tag{4.3-12}$$

The situation is illustrated for a case with $m = n = 2$, $r = 1$, and the error distribution II in Fig. 4.13. The subspace $\langle \mathbf{K} \rangle$ is just a line through the origin in b-space,

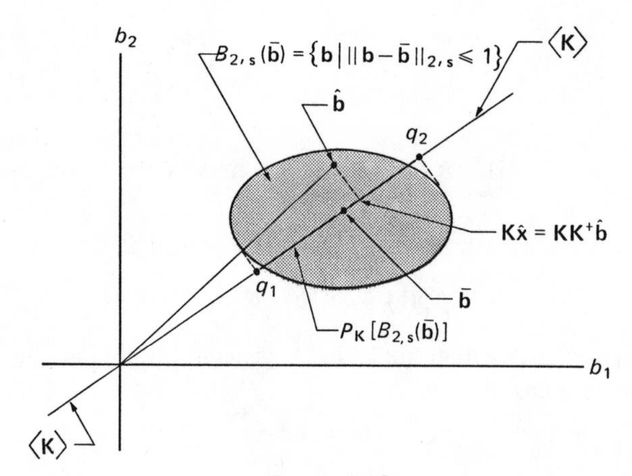

Figure 4.13.

and the entire x-space maps into this line under the mapping

$$\mathbf{b} = \mathbf{Kx}.$$

The vector $\bar{\mathbf{b}}$ is on this line and the system (4.3-11) defines a linear coset $\{\mathbf{x}(\bar{\mathbf{b}})\}$ of solutions, which is shown in Fig. 4.14.

The sample vector $\hat{\mathbf{b}}$ does not lie in the subspace $\langle \mathbf{K} \rangle$, but if $\hat{\mathbf{x}}$ is any solution to the least squares problem (4.3-11), then the vector $\mathbf{K}\hat{\mathbf{x}}$ is in $\langle \mathbf{K} \rangle$ and the residual vector $\hat{\mathbf{b}} - \mathbf{K}\hat{\mathbf{x}}$ is orthogonal to $\langle \mathbf{K} \rangle$. To see this, we note that

$$\hat{\mathbf{x}} = \mathbf{K}^\dagger \hat{\mathbf{b}} + (\mathbf{I} - \mathbf{K}^\dagger \mathbf{K})\,\mathbf{v}$$

for some n-vector $\mathbf{v}$, and since

$$\mathbf{K}\mathbf{K}^\dagger \mathbf{K} = \mathbf{K},$$

it follows that

$$\mathbf{K}\hat{\mathbf{x}} = \mathbf{K}\mathbf{K}^{\dagger}\hat{\mathbf{b}}. \tag{4.3-13}$$

Using this result and the fact that $\mathbf{KK}^{\dagger}$ is a symmetric matrix, we can write

$$(\hat{\mathbf{b}} - \mathbf{K}\hat{\mathbf{x}})^T\mathbf{K} = (\hat{\mathbf{b}} - \mathbf{K}\mathbf{K}^{\dagger}\hat{\mathbf{b}})^T\mathbf{K} = \hat{\mathbf{b}}^T\mathbf{K} - \hat{\mathbf{b}}^T\mathbf{K}\mathbf{K}^{\dagger}\mathbf{K} = \mathbf{0},$$

which means that $(\hat{\mathbf{b}} - \mathbf{K}\hat{\mathbf{x}})$ is orthogonal to all the columns of $\mathbf{K}$ and hence is orthogonal to $\langle \mathbf{K} \rangle$. Thus, $\mathbf{K}\hat{\mathbf{x}} = \mathbf{K}\mathbf{K}^{\dagger}\hat{\mathbf{b}}$ is just the orthogonal projection of $\hat{\mathbf{b}}$ into $\langle \mathbf{K} \rangle$, and the vector $\hat{\mathbf{x}}$ is a solution of the least squares problem (4.3-11) if

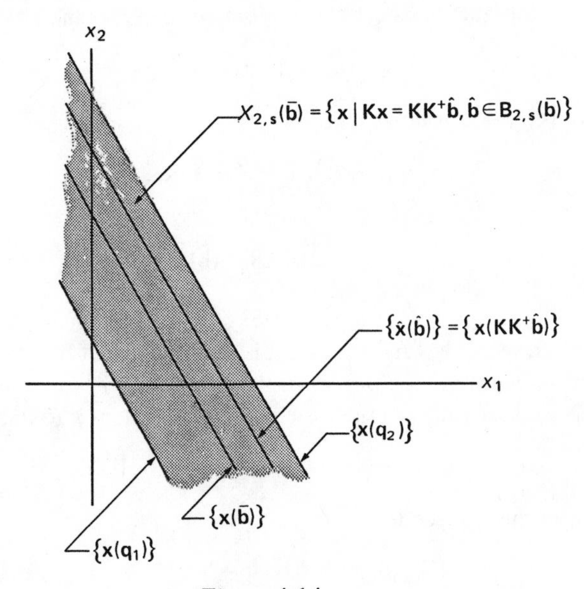

Figure 4.14.

and only if it is an exact solution of the consistent linear system (4.3-13). It follows then that the set, in x-space, of all solutions to least squares problems corresponding to $\mathbf{b}$-vectors in $B_{h,s}(\bar{\mathbf{b}})$ is just the same as the set of all exact solutions to the systems

$$\mathbf{Kx} = \mathbf{b}$$

corresponding to vectors $\mathbf{b} = \mathbf{K}\mathbf{K}^{\dagger}\mathbf{b}$ in the projection of $B_{h,s}(\bar{\mathbf{b}})$ into the subspace $\langle \mathbf{K} \rangle$. This set can be written

$$X_{h,\mathbf{s}}(\bar{\mathbf{b}}) = \{\mathbf{x} | \mathbf{Kx} = \mathbf{K}\mathbf{K}^{\dagger}\mathbf{b}, \|\mathbf{b} - \bar{\mathbf{b}}\|_{h,\mathbf{s}} \leqslant 1\}.$$

If the projection of $B_{h,s}(\bar{\mathbf{b}})$ into $\langle \mathbf{K} \rangle$ is written

$$P_{\mathbf{K}}[B_{h,s}(\bar{\mathbf{b}})] = \{\mathbf{b} | \mathbf{K}\mathbf{K}^{\dagger}\mathbf{b} = \mathbf{b} \text{ and } \|\mathbf{b} - \bar{\mathbf{b}}\|_{h,s} \leqslant 1\},$$

then $X_{h,s}(\bar{\mathbf{b}})$ can also be written

$$X_{h,s}(\bar{\mathbf{b}}) = \{\mathbf{x} | \mathbf{K}\mathbf{x} = \mathbf{b}, \mathbf{b} \in P_{\mathbf{K}}[B_{h,s}(\bar{\mathbf{b}})]\}.$$

In Fig. 4.13 the set $P_{\mathbf{K}}[B_{2,s}(\bar{\mathbf{b}})]$ is just that segment of the line $\langle \mathbf{K} \rangle$ which lies between the points (vectors) $\mathbf{q}_1$ and $\mathbf{q}_2$. The set $X_{2,s}(\bar{\mathbf{b}})$ is the shaded area in Fig. 4.14 that is bounded by the two linear cosets $\{\mathbf{x}(\mathbf{q}_1)\}$ and $\{\mathbf{x}(\mathbf{q}_2)\}$.

We do not actually know the vector $\bar{\mathbf{b}}$, so we cannot determine the sets $B_{h,s}(\bar{\mathbf{b}})$, $P_{\mathbf{K}}[B_{h,s}(\bar{\mathbf{b}})]$, and $X_{h,s}(\bar{\mathbf{b}})$. What we do know is a sample vector $\hat{\mathbf{b}}$ that was drawn from the set $B_{h,s}(\bar{\mathbf{b}})$. Therefore, we consider the set

$$B_{h,s}(\hat{\mathbf{b}}) = \{\mathbf{b} | \|\mathbf{b} - \hat{\mathbf{b}}\|_{h,s} \leqslant 1\}. \tag{4.3-14}$$

From Eq. (4.3-6), it follows that

$$\|\hat{\mathbf{b}} - \bar{\mathbf{b}}\|_{h,s} = \|\bar{\mathbf{b}} - \hat{\mathbf{b}}\|_{h,s} \leqslant 1,$$

so that

$$\bar{\mathbf{b}} \in B_{h,s}(\hat{\mathbf{b}}). \tag{4.3-15}$$

For any given sample vector $\hat{\mathbf{b}}$, the set $B_{h,s}(\hat{\mathbf{b}})$ is always guaranteed to contain the unknown vector $\bar{\mathbf{b}}$; furthermore, it is the smallest set for which this guarantee can be made. Since the vector $\bar{\mathbf{b}}$ must simultaneously be in $B_{h,s}(\hat{\mathbf{b}})$ and the subspace $\langle \mathbf{K} \rangle$, we need only consider the set of all solutions $\mathbf{x}$ to linear systems of the form

$$\mathbf{K}\mathbf{x} = \mathbf{b}$$

where $\mathbf{b}$ is in the intersection

$$\langle \mathbf{K} \rangle \cap B_{h,s}(\hat{\mathbf{b}}) = \{\mathbf{b} | \mathbf{K}\mathbf{K}^{\dagger}\mathbf{b} = \mathbf{b} \text{ and } \|\mathbf{b} - \hat{\mathbf{b}}\|_{h,s} \leqslant 1\}.$$

Thus, all of the solutions $\bar{\mathbf{x}}$ to the system (4.3-10) are guaranteed to be in the set

$$\begin{aligned} X_{h,s}(\hat{\mathbf{b}}) &= \{\mathbf{x} | \mathbf{K}\mathbf{x} = \mathbf{b}, \mathbf{b} \in \langle \mathbf{K} \rangle \cap B_{h,s}(\hat{\mathbf{b}})\} \\ &= \{\mathbf{x} | \mathbf{K}\mathbf{x} = \mathbf{b}, \mathbf{b} = \mathbf{K}\mathbf{K}^{\dagger}\mathbf{b}, \|\mathbf{b} - \hat{\mathbf{b}}\|_{h,s} \leqslant 1\}. \end{aligned}$$

The three conditions defining this set can be simultaneously expressed by the single condition

$$\|\mathbf{K}\mathbf{x} - \hat{\mathbf{b}}\|_{h,s} \leqslant 1,$$

so that $X_{h,s}(\hat{\mathbf{b}})$ can be written

$$X_{h,s}(\hat{\mathbf{b}}) = \{\mathbf{x} | \|\mathbf{K}\mathbf{x} - \hat{\mathbf{b}}\|_{h,s} \leqslant 1\}. \tag{4.3-16}$$

For a given sample $\hat{\mathbf{b}}$, this is the smallest set that can be guaranteed to contain the solutions $\bar{\mathbf{x}}$. The situation is illustrated in Fig. 4.15a, b for the case where $m=n=2, r=1$, and $h=2$.

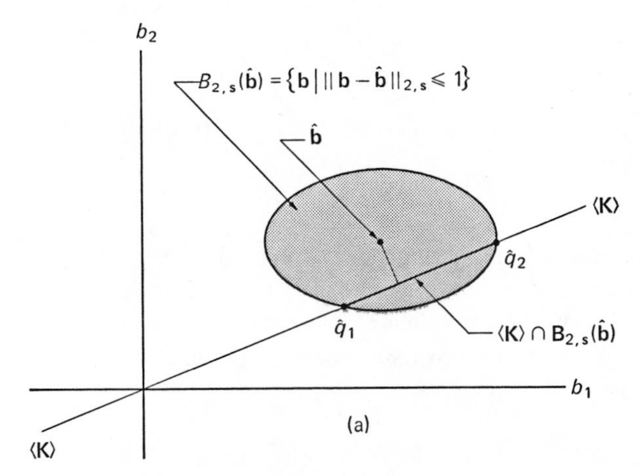

Figure 4.15a.

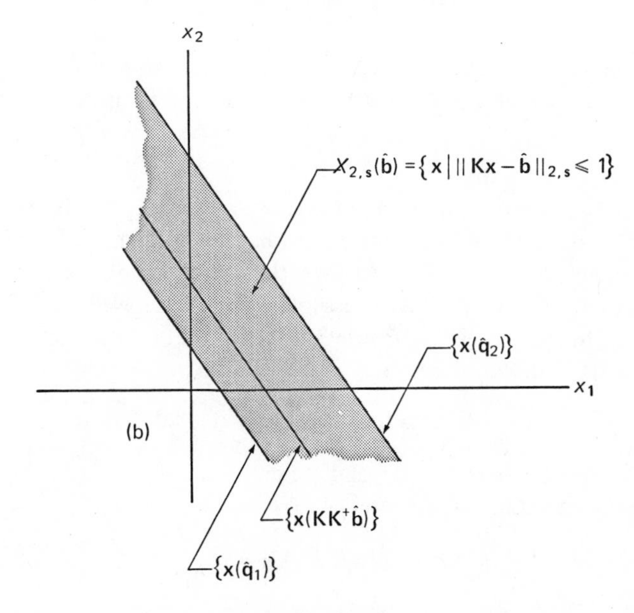

Figure 4.15b.

The final element of the problem we are considering is a given set of n-vectors $\mathbf{w}_1, \mathbf{w}_2, \ldots, \mathbf{w}_k$, which are called window vectors. The problem to be solved is to use the matrix $\mathbf{K}$ and the sample vector $\hat{\mathbf{b}}$ to find estimates for the quantities $\bar{p}_1$, $\bar{p}_2, \ldots, \bar{p}_k$, which are defined by

$$\bar{p}_i = \mathbf{w}_i^T \bar{\mathbf{x}}, \qquad i = 1, 2, \ldots, k. \tag{4.3-17}$$

Of course, for a given $\mathbf{w}_i$, there may be very many possible values of $\bar{p}_i$ corresponding to the many possible solutions $\bar{\mathbf{x}}$ of (4.3-10). We therefore seek, for each of the $\bar{p}_i$, an interval

$$I_{\hat{p}i} = [\hat{p}_i^{\,\mathrm{lo}}, \hat{p}_i^{\,\mathrm{up}}] \tag{4.3-18}$$

that is guaranteed to contain all of the possible values of $\bar{p}_i$. Since all of the $\bar{\mathbf{x}}$ are known to lie in the set $X_{h,\mathbf{s}}(\hat{\mathbf{b}})$, and since this is the smallest set for which that statement can be made with absolute certainty, it is natural to take for $\hat{p}_i^{\mathrm{lo}}$ and $\hat{p}_i^{\mathrm{up}}$ the minimum and maximum values of the function

$$\hat{p}_i = \mathbf{w}_i^T \mathbf{x} \tag{4.3-19}$$

when $\mathbf{x}$ is constrained to lie in the set $X_{h,\mathbf{s}}(\hat{\mathbf{b}})$; that is,

$$\hat{p}_i^{\,\mathrm{lo}} = \min_{\mathbf{x}} \{\mathbf{w}_i^T \mathbf{x} \mid \mathbf{x} \in X_{h,\mathbf{s}}(\hat{\mathbf{b}})\}$$

$$\hat{p}_i^{\,\mathrm{up}} = \max_{\mathbf{x}} \{\mathbf{w}_i^T \mathbf{x} \mid \mathbf{x} \in X_{h,\mathbf{s}}(\hat{\mathbf{b}})\}.$$

If the columns of $\mathbf{K}$ are linearly dependent, then in most cases the only guaranteed interval that can be obtained from the foregoing expressions is the trivial interval

$$I_{\hat{p}i} = [-\infty, \infty],$$

for if $\mathbf{w}_i$ has a nonzero projection in the subspace spanned by the eigenvectors of $\mathbf{K}^T\mathbf{K}$ that correspond to the zero singular values of $\mathbf{K}$, then the function (4.3-19) will be unbounded in the set $X_{h,\mathbf{s}}(\hat{\mathbf{b}})$. We, therefore, do not seek the minimum and maximum values that $\mathbf{w}_i^T \mathbf{x}$ can assume when $\mathbf{x}$ is allowed to vary freely in the set $X_{h,\mathbf{s}}(\hat{\mathbf{b}})$, but instead we restrict our attention to those vectors in $X_{h,\mathbf{s}}(\hat{\mathbf{b}})$ that also satisfy the constraint that

$$x_i \geqslant 0, \qquad i = 1, 2, \ldots, n. \tag{4.3-20}$$

Using this nonnegative constraint, and the expression (4.3-16) for $X_{h,\mathbf{s}}(\hat{\mathbf{b}})$, we can write the end points of the interval $I_{\hat{p}_i}$ as

$$\hat{p}_i^{\,\mathrm{lo}} = \min_{\mathbf{x} \geqslant \mathbf{0}} \{\mathbf{w}_i^T \mathbf{x} \mid \|\mathbf{Kx} - \hat{\mathbf{b}}\|_{h,\mathbf{s}} \leqslant 1\}, \tag{4.3-21}$$

$$\hat{p}_i^{\,\mathrm{up}} = \max_{\mathbf{x} \geqslant \mathbf{0}} \{\mathbf{w}_i^T \mathbf{x} \mid \|\mathbf{Kx} - \hat{\mathbf{b}}\|_{h,\mathbf{s}} \leqslant 1\}. \tag{4.3-22}$$

Since $\mathbf{w}_i^T$, $\mathbf{K}$, $\hat{\mathbf{b}}$, h, and s are all known quantities, the interval $I_{\hat{p}_i}$ is then, in principle at least, a calculable or obtainable interval; *furthermore, it is the narrowest possible interval that can be obtained from the information that is given.*

The problem of obtaining the interval $I_{\hat{p}_i}$ has now been reduced to a constrained minimization and a constrained maximization problem. It is easy to see that, for each of the three kinds of error distributions that concern us here, the problem becomes either a linear programming or a quadratic programming problem. For error distribution I, Eq. (4.3-3), the limits of the interval are given by

$$\hat{p}_i^{\mathrm{lo}} = \min_{\mathbf{x} \geqslant \mathbf{o}} \left\{ \mathbf{w}_i^T \mathbf{x} \,\middle|\, \sum_{j=1}^{m} \frac{|(\mathbf{K}\mathbf{x} - \hat{\mathbf{b}})_j|}{s_j} \leqslant 1 \right\} \tag{4.3-23}$$

and

$$\hat{p}_i^{\mathrm{up}} = \max_{\mathbf{x} \geqslant \mathbf{o}} \left\{ \mathbf{w}_i^T \mathbf{x} \,\middle|\, \sum_{j=1}^{m} \frac{|(\mathbf{K}\mathbf{x} - \hat{\mathbf{b}})_j|}{s_j} \leqslant 1 \right\}. \tag{4.3-24}$$

Although this is not the standard way of stating linear programming problems, it is clear that, in each case, what is sought is the extreme value of a linear function of $\mathbf{x}$ when $\mathbf{x}$ is subject to a set of linear constraints that includes the constraint that all the elements of $\mathbf{x}$ are nonnegative, and this is just a general statement of the linear programming problem. For the error distribution II, Eq. (4.3-4), the limits of the interval become

$$\hat{p}_i^{\mathrm{lo}} = \min_{\mathbf{x} \geqslant \mathbf{o}} \left\{ \mathbf{w}_i^T \mathbf{x} \,\middle|\, \left(\sum_{j=1}^{m} \frac{|(\mathbf{K}\mathbf{x} - \hat{\mathbf{b}})_j|^2}{s_j^2} \right)^{1/2} \leqslant 1 \right\} \tag{4.3-25}$$

and

$$\hat{p}_i^{\mathrm{up}} = \max_{\mathbf{x} \geqslant \mathbf{o}} \left\{ \mathbf{w}_i^T \mathbf{x} \,\middle|\, \left(\sum_{j=1}^{m} \frac{|(\mathbf{K}\mathbf{x} - \hat{\mathbf{b}})_j|^2}{s_j^2} \right)^{1/2} \leqslant 1 \right\}. \tag{4.3-26}$$

Thus, the problem is one of finding the extreme values of a linear function subject to quadratic constraints and a nonnegativity constraint, and hence it is a quadratic programming problem. Finally, for error distribution III, the limits are

$$\hat{p}_i^{\mathrm{lo}} = \min_{\mathbf{x} \geqslant \mathbf{o}} \left\{ \mathbf{w}_i^T \mathbf{x} \,\middle|\, \max_{1 \leqslant j \leqslant m} \frac{|(\mathbf{K}\mathbf{x} - \hat{\mathbf{b}})_j|}{s_j} \leqslant 1 \right\}, \tag{4.3-27}$$

and

$$\hat{p}_i^{\text{up}} = \max_{\mathbf{x} \geqslant \mathbf{o}} \left\{ \mathbf{w}_i^T \mathbf{x} \,\middle|\, \max_{1 \leqslant j \leqslant m} \frac{|(\mathbf{K}\mathbf{x} - \hat{\mathbf{b}})_j|}{s_j} \leqslant 1 \right\}. \tag{4.3-28}$$

Both of these optimizations are linear programming problems, and it is quite easy to restate them in the standard linear programming form. To see this, note that the constraint

$$\max_{1 \leqslant j \leqslant m} \frac{|(\mathbf{K}\mathbf{x} - \hat{\mathbf{b}})_j|}{s_j} \leqslant 1$$

can also be written

$$|(\mathbf{K}\mathbf{x} - \hat{\mathbf{b}})_j| \leqslant s_j, \qquad j = 1, 2, \ldots, m,$$

which is the same as

$$-s_j \leqslant (\mathbf{K}\mathbf{x} - \hat{\mathbf{b}})_j \leqslant s_j, \qquad j = 1, 2, \ldots, m.$$

This last condition can be replaced by the two inequalities

$$(\mathbf{K}\mathbf{x})_j \leqslant \hat{b}_j + s_j, \qquad j = 1, 2, \ldots, m,$$

and

$$-(\mathbf{K}\mathbf{x})_j \leqslant -\hat{b}_j + s_j, \qquad j = 1, 2, \ldots, m,$$

and these two inequalities, when expressed in matrix form, become

$$\mathbf{K}\mathbf{x} \leqslant \hat{\mathbf{b}} + \mathbf{s}', \qquad -\mathbf{K}\mathbf{x} \leqslant -\hat{\mathbf{b}} + \mathbf{s}'$$

where $\mathbf{s}' = (s_1, s_2, \ldots, s_m)^T$.

Therefore, if $\mathbf{A}$ is the $2m \times n$ matrix defined by

$$\mathbf{A} = \begin{pmatrix} \mathbf{K} \\ -\mathbf{K} \end{pmatrix}, \tag{4.3-29}$$

and $\mathbf{c}$ is the $2m$-vector

$$\mathbf{c} = \begin{pmatrix} \hat{\mathbf{b}} + \mathbf{s}' \\ -\hat{\mathbf{b}} + \mathbf{s}' \end{pmatrix}, \tag{4.3-30}$$

then the constraint can be written

$$\mathbf{A}\mathbf{x} \leqslant \mathbf{c} \tag{4.3-31}$$

and the end points of the interval can be written

$$\hat{p}_i^{\text{lo}} = \min_{\mathbf{x} \geqslant \mathbf{o}} \{\mathbf{w}_i^T \mathbf{x} | \mathbf{A}\mathbf{x} \leqslant \mathbf{c}\} \tag{4.3-32}$$

and

$$\hat{p}_i^{\text{up}} = \max_{\mathbf{x} \geqslant \mathbf{0}} \{\mathbf{w}_i^T \mathbf{x} | \mathbf{A}\mathbf{x} \leqslant \mathbf{c}\}. \tag{4.3-33}$$

These last two equations are standard ways of stating linear programming problems. Similar, though somewhat more involved, arguments can be used to reduce Eqs. (4.3-23) and (4.3-24) to standard linear programming representations, and likewise (4.3-25) and (4.3-26) to standard quadratic programming representations.

Although we have suceeded in stating our constrained estimation problems as problems in mathematical programming, it should be noted that, for many particular problems, solutions will not exist. Two situations can arise in which it will not be possible to obtain a nontrivial solution. We have seen that, for an unconstrained underdetermined problem, the region of possible solutions will be unbounded in the directions of the eigenvectors of $\mathbf{K}^T\mathbf{K}$ corresponding to zero singular values of $\mathbf{K}$, and most window vectors will have nonzero components in these directions, so that no nontrivial interval for the corresponding window function can be found. The constrained estimation problem, and hence the corresponding mathematical programming problem, will have a solution in these cases only when the intersection of the set of solution vectors with the positive orthant is bounded in the direction of the window vector. If the window vector has a component in any direction in which the constrained solution set is unbounded, then the corresponding mathematical programming problem has an unbounded solution. The other situation in which no solution can be found occurs when the unconstrained solution set does not have any points in common with the positive orthant. In cases like this the constraints in the corresponding mathematical programming problem are inconsistent and no solution exists, not even a trivial unbounded solution.

Readers who are not very familiar with the field of mathematical programming may want to consult some introductory texts for more details about the formulation and solution of mathematical programming problems. Two excellent texts for this purpose are Gass's *Linear Programming* [1], and Kunzi, Krelle, and Oettli's *Nonlinear Programming* [2].

4.4 A Dual Formulation of the Problem

In the preceding section we showed that constrained linear estimation is a problem in mathematical programming. One of the most basic relations of mathematical programming is a duality theorem that allows any given problem to be stated in two different, but equivalent, ways–one way as a constrained maximization problem and the other way as a constrained minimization problem.

One way of stating the problem is called the *primal,* and the other way is called the *dual* (but which is called which is a matter of convention).

If the solution space of the primal in an N-dimensional space with unknown variables $z_1, z_2, \ldots, z_N$ and the constraint space (the b-space in our problems) is an M-dimensional space, then the solution space of the dual will be M-dimensional with unknown variables $v_1, v_2, \ldots, v_M$ and the constraint space in the dual will be N-dimensional.

A statement of this duality theorem has been given by Wolfe [3], who adopts the convention that the primal is a minimization problem and that the dual is a maximization problem. Let $\mathbf{z}$ and $\mathbf{v}$ be the vectors of the primal and dual variables, that is,

$$\mathbf{z} = \begin{bmatrix} z_1 \\ z_2 \\ \cdot \\ \cdot \\ \cdot \\ z_N \end{bmatrix}, \qquad \mathbf{v} = \begin{bmatrix} v_1 \\ v_2 \\ \cdot \\ \cdot \\ \cdot \\ v_M \end{bmatrix}. \tag{4.4.-1}$$

Then in Wolfe's statement of the theorem, the primal problem is

$$P: \phi_P = \min_{\mathbf{z}} \{f(\mathbf{z}) | \; g_i(\mathbf{z}) \geqslant 0, i = 1, 2, \ldots, M\}, \tag{4.4-2}$$

that is, to minimize a given function $f(\mathbf{z})$ of the vector $\mathbf{z}$ subject to the M constraints $g_i(\mathbf{z}) \geqslant 0$. The dual problem is

$$D: \phi_D = \max_{(\mathbf{z},\mathbf{v})} \{ [f(\mathbf{z}) - \sum_{i=1}^{M} v_i g_i(\mathbf{z})] | \nabla f(\mathbf{z}) = \sum_{i=1}^{M} v_i \nabla g_i(\mathbf{z}) \text{ and } \mathbf{v} \geqslant \mathbf{0}\} \tag{4.4-3}$$

where ∇ is the gradient operator, that is,

$$\nabla = \begin{bmatrix} \dfrac{\partial}{\partial z_1} \\ \dfrac{\partial}{\partial z_2} \\ \vdots \\ \dfrac{\partial}{\partial z_N} \end{bmatrix}.$$

The theorem states that if there exists a vector $\mathbf{z}^*$ that solves the primal (gives a finite minimum ϕ_P), then there exists a vector $\mathbf{v}^*$ such that $\mathbf{z}^*$ and $\mathbf{v}^*$ together solve the dual, and that the two extreme values are equal, that is,

$$\phi_P = \phi_D. \tag{4.4-4}$$

If the vector $\mathbf{g}$ is defined by

$$\mathbf{g} = \begin{bmatrix} g_1(\mathbf{z}) \\ g_2(\mathbf{z}) \\ \vdots \\ g_M(\mathbf{z}) \end{bmatrix} \tag{4.4.-5}$$

and if $\partial \mathbf{g}^T/\partial \mathbf{z}$ is the matrix defined by

$$\frac{\partial \mathbf{g}^T}{\partial \mathbf{z}} = \frac{\partial}{\partial \mathbf{z}}(g_1(\mathbf{z}), g_2(\mathbf{z}), \ldots, g_M(\mathbf{z}))$$

$$= \left(\frac{\partial}{\partial \mathbf{z}} g_1(\mathbf{z}), \frac{\partial}{\partial \mathbf{z}} g_2(\mathbf{z}), \ldots, \frac{\partial}{\partial \mathbf{z}} g_M(\mathbf{z})\right) \tag{4.4-6}$$

or

$$\frac{\partial \mathbf{g}^T}{\partial \mathbf{z}} = \begin{bmatrix} \frac{\partial g_1}{\partial z_1} & \frac{\partial g_2}{\partial z_1} & \cdots & \frac{\partial g_M}{\partial z_1} \\ \frac{\partial g_1}{\partial z_2} & \frac{\partial g_2}{\partial z_2} & \cdots & \frac{\partial g_M}{\partial z_2} \\ \vdots & \vdots & & \vdots \\ \frac{\partial g_1}{\partial z_M} & \frac{\partial g_2}{\partial z_M} & \cdots & \frac{\partial g_M}{\partial z_M} \end{bmatrix}, \tag{4.4-7}$$

then the primal problem can be stated as

$$P\colon\ \phi_P = \min_{\mathbf{z}} \{f(\mathbf{z}) | \mathbf{g} \geqslant 0\}, \tag{4.4-8}$$

and the dual problem can be written

$$D\colon\ \phi_D = \max_{(\mathbf{z},\mathbf{v})} \{f(\mathbf{z}) - \mathbf{g}^T \mathbf{v} | \nabla f(\mathbf{z}) = \left(\frac{\partial \mathbf{g}^T}{\partial \mathbf{z}}\right)\mathbf{v}, \mathbf{v} \geqslant \mathbf{0}\}. \tag{4.4-9}$$

We can now apply the Wolfe duality theorem to the mathematical programming problems in the previous section. In particular, consider the quadratic programming problem that arises from error distribution II, Eq. (4.3-4). According to Eq. (4.3-25), the left-hand end point of the interval $I_{\hat{p}_i}$ is

$$\hat{p}_i^{\text{lo}} = \min_{\mathbf{x} \geqslant \mathbf{0}} \left\{ \mathbf{w}_i^T \mathbf{x} \left| \left[\sum_{j=1}^{m} \frac{|\mathbf{Kx} - \hat{\mathbf{b}}|_j^2}{s_j^2} \right]^{1/2} \leqslant 1 \right. \right\}. \tag{4.4-10}$$

In order to apply the duality theorem, we restate the problem in a slightly different but equivalent form. The constraint

$$\left[\sum_{j=1}^{m} \frac{|\mathbf{Kx} - \hat{\mathbf{b}}|_j^2}{s_j^2} \right]^{1/2} \leqslant 1$$

can also be written

$$(\mathbf{Kx} - \hat{\mathbf{b}})^T \mathbf{S}^{-2} (\mathbf{Kx} - \hat{\mathbf{b}}) \leqslant 1$$

where $\mathbf{S}^{-1}$ is the diagonal matrix defined by

$$\mathbf{S}^{-1} = \operatorname{diag}\left(\frac{1}{s_1}, \frac{1}{s_2}, \ldots, \frac{1}{s_m} \right) \tag{4.4-11}$$

and both sides of the inequality have been squared. Thus the primal problem can be written

$$\hat{p}_i^{\text{lo}} = \min \{ \mathbf{w}_i^T \mathbf{x} | (\mathbf{Kx} - \hat{\mathbf{b}})^T \mathbf{S}^{-2} (\mathbf{Kx} - \hat{\mathbf{b}}) \leqslant 1, \mathbf{x} \geqslant \mathbf{0} \}. \tag{4.4-12}$$

We now define a new vector $\mathbf{r}$ by

$$\mathbf{r} = \mathbf{S}^{-1} (\mathbf{Kx} - \hat{\mathbf{b}}), \tag{4.4-13}$$

or equivalently by

$$\mathbf{r} \leqslant \mathbf{S}^{-1} (\mathbf{Kx} - \hat{\mathbf{b}}) \leqslant \mathbf{r},$$

from which it follows that

$$\mathbf{r} - \mathbf{S}^{-1} (\mathbf{Kx} - \hat{\mathbf{b}}) \geqslant \mathbf{0} \tag{4.4-14}$$

and

$$-\mathbf{r} + \mathbf{S}^{-1} (\mathbf{Kx} - \hat{\mathbf{b}}) \geqslant \mathbf{0}. \tag{4.4-15}$$

In terms of $\mathbf{r}$, the constraint

$$(\mathbf{Kx} - \hat{\mathbf{b}})^T \mathbf{S}^{-2} (\mathbf{Kx} - \hat{\mathbf{b}}) \leqslant 1$$

can be written

$$1 - \mathbf{r}^T \mathbf{r} \geqslant 0. \tag{4.4-16}$$

Using the constraint expressions (4.4-14)-(4.4-16), the primal problem (4.4-12) can be rewritten as

$$\hat{p}_i^{\,lo} = \min_{(\mathbf{x},\mathbf{r})} \{(\mathbf{w}_i^T, 0) \begin{pmatrix} \mathbf{x} \\ \mathbf{r} \end{pmatrix} | \mathbf{r} - \mathbf{S}^{-1}(\mathbf{K}\mathbf{x} - \hat{\mathbf{b}}) \geqslant \mathbf{0}, -\mathbf{r} + \mathbf{S}^{-1}(\mathbf{K}\mathbf{x} - \hat{\mathbf{b}}) \geqslant \mathbf{0}$$

$$1 - \mathbf{r}^T \mathbf{r} \geqslant 0, \mathbf{x} \geqslant \mathbf{0}\}, \tag{4.4-17}$$

which is a form convenient for the application of the duality theorem. Let $\mathbf{z}$ be the $(n+m)$-vector

$$\mathbf{z} = \begin{pmatrix} \mathbf{x} \\ \mathbf{r} \end{pmatrix} \tag{4.4-18}$$

and $\mathbf{g}$ the $(2m+n+1)$-vector

$$\mathbf{g} = \begin{pmatrix} \mathbf{r} - \mathbf{S}^{-1}(\mathbf{K}\mathbf{x} - \hat{\mathbf{b}}) \\ -\mathbf{r} + \mathbf{S}^{-1}(\mathbf{K}\mathbf{x} - \hat{\mathbf{b}}) \\ \mathbf{x} \\ 1 - \mathbf{r}^T \mathbf{r} \end{pmatrix}. \tag{4.4-19}$$

The function $f(\mathbf{z})$ is given by

$$f(\mathbf{z}) = (\mathbf{w}_i^T, 0)\, \mathbf{z} \tag{4.4-20}$$

so that $\nabla f(\mathbf{z})$ is the $(n+m)$-vector

$$\nabla f(\mathbf{z}) = \begin{pmatrix} \mathbf{w}_i \\ \mathbf{0} \end{pmatrix}. \tag{4.4-21}$$

The matrix $(\partial \mathbf{g}^T / \partial \mathbf{z})$ is the $(n+m) \times (2m+n+1)$ matrix

$$\frac{\partial \mathbf{g}^T}{\partial \mathbf{z}} = \begin{bmatrix} -\mathbf{K}^T \mathbf{S}^{-1} & \mathbf{K}^T \mathbf{S}^{-1} & \mathbf{I}_n & \mathbf{0} \\ \mathbf{I}_m & -\mathbf{I}_m & \mathbf{0} & -2\mathbf{r} \end{bmatrix} \tag{4.4-22}$$

where $\mathbf{I}_m$ and $\mathbf{I}_n$ denote the identity matrices of order m and n. The dual unknown vector will be a $(2m+n+1)$-vector. We write it in the partitioned form

$$\mathbf{v} = \begin{pmatrix} \mathbf{u}^- \\ \mathbf{u}^+ \\ \mathbf{y} \\ \beta \end{pmatrix} \tag{4.4-23}$$

where $\mathbf{u}^+$ and $\mathbf{u}^-$ are both m-vectors, $\mathbf{y}$ is an n-vector, and β is a scalar. Since the dual vector is constrained to be nonnegative, we have

$$\mathbf{u}^+ \geqslant \mathbf{0}, \quad \mathbf{u}^- \geqslant \mathbf{0}, \quad \mathbf{y} \geqslant \mathbf{0}, \quad \beta \geqslant 0. \tag{4.4-24}$$

The other constraint in the dual problem is

$$\nabla f(\mathbf{z}) = \left(\frac{\partial \mathbf{g}^T}{\partial \mathbf{z}}\right)\mathbf{v}$$

and substituting Eqs. (4.4-21), (4.4-22), (4.4-23) into this expression gives

$$\mathbf{w}_i = -\mathbf{K}^T\mathbf{S}^{-1}\mathbf{u}^- + \mathbf{K}^T\mathbf{S}^{-1}\mathbf{u}^+ + \mathbf{y}$$

and

$$\mathbf{0} = \mathbf{u}^- - \mathbf{u}^+ - 2\beta\mathbf{r}.$$

If we define the vector $\mathbf{u}$ by

$$\mathbf{u} = \mathbf{u}^+ - \mathbf{u}^-, \tag{4.4-25}$$

then these last two constraint equations can be written

$$\mathbf{w}_i = \mathbf{K}^T\mathbf{S}^{-1}\mathbf{u} + \mathbf{y} \tag{4.4-26}$$

and

$$\mathbf{r} = -\frac{1}{2\beta}\mathbf{u}. \tag{4.4-27}$$

From Eqs. (4.4-19), (4.4-20), (4.4-23), and (4.4-25), it follows that the objective function for the dual problem (the function to be maximized) can be written

$$f(\mathbf{z}) - \mathbf{g}^T\mathbf{v} = \mathbf{w}_i^T\mathbf{x} - [-\mathbf{r}^T\mathbf{u} + \mathbf{x}^T\mathbf{K}^T\mathbf{S}^{-1}\mathbf{u} - \hat{\mathbf{b}}^T\mathbf{S}^{-1}\mathbf{u} + \mathbf{x}^T\mathbf{y} + \beta(1 - \mathbf{r}^T\mathbf{r})].$$

Substitution of the constraint relations (4.4-26) and (4.4-27) into this expression reduces it to

$$f(\mathbf{z}) - \mathbf{g}^T\mathbf{v} = \hat{\mathbf{b}}^T\mathbf{S}^{-1}\mathbf{u} - \beta[1 + (4\beta^2)^{-1}\mathbf{u}^T\mathbf{u}]. \tag{4.4-28}$$

Using expressions (4.4-24)-(4.4-28), we can write the dual problem in the form

$$\begin{aligned}\hat{p}_i^{lo} = \max_{(\mathbf{x},\mathbf{r},\mathbf{u},\mathbf{y},\beta)} \{&\hat{\mathbf{b}}^T\mathbf{S}^{-1}\mathbf{u} - \beta[1 + (4\beta^2)^{-1}\mathbf{u}^T\mathbf{u}] | \mathbf{K}^T\mathbf{S}^{-1}\mathbf{u} \\ &= \mathbf{w}_i - \mathbf{y}, \mathbf{r} = -(2\beta)^{-1}\mathbf{u}, \\ &\mathbf{u} = \mathbf{u}^+ - \mathbf{u}^-, \mathbf{u}^+ \geqslant \mathbf{0}, \mathbf{u}^- \geqslant \mathbf{0}, \mathbf{y} \geqslant \mathbf{0}, \beta \geqslant 0\}.\end{aligned} \tag{4.4-29}$$

Now since $\mathbf{y}$ does not appear in the objective function, we can write the two constraints

$$\mathbf{K}^T\mathbf{S}^{-1}\mathbf{u} = \mathbf{w}_i - y, \qquad y \geq 0,$$

more simply as

$$\mathbf{K}^T\mathbf{S}^{-1}\mathbf{u} \leq \mathbf{w}_i.$$

The variables $\mathbf{x}$, $\mathbf{u}^+$, $\mathbf{u}^-$, and $\mathbf{r}$ do not appear in the objective function either. The $\mathbf{x}$ was eliminated in Eq. (4.4-28); $\mathbf{u}^+$ is just a vector whose nonzero elements are the nonnegative elements of the unconstrained vector $\mathbf{u}$; $\mathbf{u}^-$ is a vector whose nonzero elements are the magnitudes of the negative elements of $\mathbf{u}$; and the vector $\mathbf{r}$ is completely determined by $\mathbf{u}$ and β. Taking all this into account, we rewrite (4.4-29) in the form

$$\hat{p}_i^{\mathrm{lo}} = \max_{(\mathbf{u},\beta)} \{\hat{\mathbf{b}}^T\mathbf{S}^{-1}\mathbf{u} - \beta[1 + (4\beta^2)^{-1}\mathbf{u}^T\mathbf{u}] | \mathbf{K}^T\mathbf{S}^{-1}\mathbf{u} \leq \mathbf{w}_i,\ \beta \geq 0\}. \tag{4.4-30}$$

Now since β is constrained to be nonnegative, the term

$$\psi(\beta) = \beta[1 + (4\beta^2)^{1}\mathbf{u}^T\mathbf{u}]$$

is nonnegative and so for any vector $\mathbf{u}$ the objective function can attain its maximum value only if $\psi(\beta)$ is minimized with respect to β. We, therefore, require that

$$\frac{d\psi}{d\beta} = 1 - (4\beta^2)^{-1}\mathbf{u}^T\mathbf{u} = 0,$$

which gives

$$\beta = \tfrac{1}{2}(\mathbf{u}^T\mathbf{u})^{1/2},$$

which is positive for any nonzero vector $\mathbf{u}$, so that the constraint $\beta \geq 0$ is satisfied. The second derivative

$$\frac{d^2\psi}{d\beta^2} = (2\beta^3)^{-1}\mathbf{u}^T\mathbf{u}$$

is positive, too, so that the extremum is a minimum. The minimum value is given by

$$\psi[\tfrac{1}{2}(\mathbf{u}^T\mathbf{u})^{1/2}] = (\mathbf{u}^T\mathbf{u})^{1/2}.$$

Thus we can rewrite the dual problem as

$$\hat{p}_i^{\mathrm{lo}} = \max_{\mathbf{u}} \{\hat{\mathbf{b}}^T\mathbf{S}^{-1}\mathbf{u} - (\mathbf{u}^T\mathbf{u})^{1/2} | \mathbf{K}^T\mathbf{S}^{-1}\mathbf{u} \leq \mathbf{w}_i\}$$

or, equivalently, as

$$\hat{p}_i^{\text{lo}} = \max_{\mathbf{u}} \{\mathbf{u}^T \mathbf{S}^{-1} \hat{\mathbf{b}} - (\mathbf{u}^T \mathbf{u})^{1/2} | \mathbf{u}^T \mathbf{S}^{-1} \mathbf{K} \leqslant \mathbf{w}_i^T\}.$$

If we define a new variable $\mathbf{u}'$ by

$$\mathbf{u}' = \mathbf{S}^{-1} \mathbf{u},$$

then the problem can be written

$$\hat{p}_i^{\text{lo}} = \max_{\mathbf{u}'} \{\mathbf{u}'^T \hat{\mathbf{b}} - (\mathbf{u}'^T \mathbf{S}^2 \mathbf{u}')^{1/2} | \mathbf{u}'^T \mathbf{K} \leqslant \mathbf{w}_i^T\}.$$

Since $\mathbf{u}'$ is simply an unconstrained independent variable, we can replace it by $\mathbf{u}'$ and write

$$\hat{p}_i^{\text{lo}} = \max_{\mathbf{u}} \{\mathbf{u}^T \hat{\mathbf{b}} - (\mathbf{u}^T \mathbf{S}^2 \mathbf{u})^{1/2} | \mathbf{u}^T \mathbf{K} \leqslant \mathbf{w}_i^T\}. \tag{4.4-31}$$

A dual statement of the problem for the upper limit of the interval $I_{\hat{p}_i}$ follows easily from what we have just done. By Eq. (4.3-26) the upper limit is

$$\hat{p}_i^{\text{up}} = \max_{\mathbf{x} \geqslant \mathbf{o}} \left\{ \mathbf{w}_i^T \mathbf{x} \middle| \left(\sum_{j=1}^{m} \frac{|\mathbf{K}\mathbf{x} - \hat{\mathbf{b}}|_j^2}{s_j^2} \right)^{1/2} \leqslant 1 \right\},$$

which is the same as

$$\hat{p}_i^{\text{up}} = \max_{\mathbf{x} \geqslant \mathbf{o}} \{\mathbf{w}_i^T \mathbf{x} | (\mathbf{K}\mathbf{x} - \hat{\mathbf{b}})^T \mathbf{S}^{-2} (\mathbf{K}\mathbf{x} - \hat{\mathbf{b}}) \leqslant 1\}. \tag{4.4-32}$$

This statement is equivalent to the statement

$$\hat{p}_i^{\text{up}} = -\min_{\mathbf{x} \geqslant \mathbf{o}} \{(-\mathbf{w}_i^T) \mathbf{x} | (\mathbf{K}\mathbf{x} - \hat{\mathbf{b}})^T \mathbf{S}^{-2} (\mathbf{K}\mathbf{x} - \hat{\mathbf{b}}) \leqslant 1\},$$

which is analogous to Eq. (4.4-12). When Wolfe's duality theorem is applied to it, the result is

$$\hat{p}_i^{\text{up}} = -\max_{\mathbf{u}} \{\mathbf{u}^T \hat{\mathbf{b}} - (\mathbf{u}^T \mathbf{S}^2 \mathbf{u})^{1/2} | \mathbf{u}^T \mathbf{K} \leqslant (-\mathbf{w}_i)^T\}, \tag{4.4-33}$$

which is analogous to Eq. (4.4-31). Since $\mathbf{u}$ is a dummy variable, we can make the substitution

$$\mathbf{u}' = -\mathbf{u}$$

and write

$$\hat{p}_i^{\text{up}} = -\max_{\mathbf{u}'} \{-\mathbf{u}'^T \hat{\mathbf{b}} - (\mathbf{u}'^T \mathbf{S}^2 \mathbf{u}')^{1/2} | -\mathbf{u}'^T \mathbf{K} \leqslant -\mathbf{w}_i^T\},$$

which is the same as

$$\hat{p}_i^{\text{up}} = \min_{\mathbf{u}} \{\mathbf{u}^T \hat{\mathbf{b}} + (\mathbf{u}^T \mathbf{S}^2 \mathbf{u})^{1/2} | \mathbf{u}^T \mathbf{K} \geqslant \mathbf{w}_i^T\}. \tag{4.4-34}$$

The linear programming problems that correspond to error distributions I and III also have dual formulations that can be obtained by arguments similar to the preceding ones. A summary of the primal and dual statements of the problem for all three types of error distributions is given in Table 4.1. Table 4.2 restates the primal and dual problems for the lower limit in each case using the notation of the vector norms that were defined by Eqs. (4.3-2)-(4.3-5). These three primal-dual relationships are special cases of a general theorem that has been proved by Charles Schneeberger [4]. According to Schneeberger's theorem, if the primal problem is

$$\phi = \min_{\mathbf{x} \geqslant \mathbf{0}} \{\mathbf{w}^T \mathbf{x} | \, \|\mathbf{K}\mathbf{x} - \mathbf{b}\|_{h,\mathbf{s}} \leqslant \mu\} \tag{4.4-35}$$

where h is an integer between 1 and ∞, $\mathbf{s}$ is a vector of the form

$$\mathbf{s}^T = \left(\frac{1}{s_1}, \frac{1}{s_2}, \ldots, \frac{1}{s_m}\right),$$

and μ is a constant, then the corresponding dual problem is

$$\phi = \max_{\mathbf{u}} \{\mathbf{u}^T \mathbf{b} - \mu \|\mathbf{u}\|_{h', \mathbf{s}'} | \mathbf{u}^T \mathbf{K} \leqslant \mathbf{w}^T\} \tag{4.4-36}$$

where

$$\mathbf{s}'^T = (s_1, s_2, \ldots, s_m)$$

and

$$\frac{1}{h} + \frac{1}{h'} = 1. \tag{4.4-37}$$

The relationship between the primal and dual formulations in the case of error distribution II ($h = h' = 2$) is analogous to the relationship between the Gauss and Markov formulations of the classical linear estimation problem. In Sections 2.2 and 2.3, we saw that if $\hat{\mathbf{b}}$ is a random vector satisfying

$$\hat{\mathbf{b}} = \mathbf{K}\bar{\mathbf{x}} + \hat{\boldsymbol{\epsilon}} \tag{4.4-38}$$

where $\hat{\boldsymbol{\epsilon}}$ is a random vector such that

$$E(\hat{\boldsymbol{\epsilon}}) = \mathbf{0}, \qquad E(\hat{\boldsymbol{\epsilon}}\hat{\boldsymbol{\epsilon}}^T) = \mathbf{S}^2, \tag{4.4-39}$$

Table 4.1

Error Distribution	Primal	Dual
I	$\hat{p}_i^{\,\mathrm{lo}} = \min\limits_{\mathbf{x} \geqslant \mathbf{o}} \left\{ \mathbf{w}_i^T \mathbf{x} \,\middle\vert\, \sum\limits_{j=1}^{m} \frac{\lvert \mathbf{Kx} - \hat{\mathbf{b}} \rvert_j}{s_j} \leqslant 1 \right\}$	$\hat{p}_i^{\,\mathrm{lo}} = \max\limits_{\mathbf{u}} \{ \mathbf{u}^T \hat{\mathbf{b}} - \max\limits_{j=1,m} [\lvert u_j \rvert s_j] \,\vert\, \mathbf{u}^T \mathbf{K} \leqslant \mathbf{w}_i^T \}$
	$\hat{p}_i^{\,\mathrm{up}} = \max\limits_{\mathbf{x} \geqslant \mathbf{o}} \left\{ \mathbf{w}_i^T \mathbf{x} \,\middle\vert\, \sum\limits_{j=1}^{m} \frac{\lvert \mathbf{Kx} - \hat{\mathbf{b}} \rvert_j}{s_j} \leqslant 1 \right\}$	$\hat{p}_i^{\,\mathrm{up}} = \min\limits_{\mathbf{u}} \{ \mathbf{u}^T \hat{\mathbf{b}} + \max\limits_{j=1,m} [\lvert u_j \rvert s_j] \,\vert\, \mathbf{u}^T \mathbf{K} \geqslant \mathbf{w}_i^T \}$
II	$\hat{p}_i^{\,\mathrm{lo}} = \min\limits_{\mathbf{x} \geqslant \mathbf{o}} \{ \mathbf{w}_i^T \mathbf{x} \,\vert\, [(\mathbf{Kx} - \hat{\mathbf{b}})^T \mathbf{S}^{-2} (\mathbf{Kx} - \hat{\mathbf{b}})] \leqslant 1 \}$	$\hat{p}_i^{\,\mathrm{lo}} = \max\limits_{\mathbf{u}} \{ \mathbf{u}^T \hat{\mathbf{b}} - (\mathbf{u}^T \mathbf{S}^2 \mathbf{u})^{1/2} \,\vert\, \mathbf{u}^T \mathbf{K} \leqslant \mathbf{w}_i^T \}$
	$\hat{p}_i^{\,\mathrm{up}} = \max\limits_{\mathbf{x} \geqslant \mathbf{o}} \{ \mathbf{w}_i^T \mathbf{x} \,\vert\, [(\mathbf{Kx} - \hat{\mathbf{b}})^T \mathbf{S}^{-2} (\mathbf{Kx} - \hat{\mathbf{b}})] \leqslant 1 \}$	$\hat{p}_i^{\,\mathrm{up}} = \min\limits_{\mathbf{u}} \{ \mathbf{u}^T \hat{\mathbf{b}} + (\mathbf{u}^T \mathbf{S}^2 \mathbf{u})^{1/2} \,\vert\, \mathbf{u}^T \mathbf{K} \geqslant \mathbf{w}_i^T \}$
III	$\hat{p}_i^{\,\mathrm{lo}} = \min\limits_{\mathbf{x} \geqslant \mathbf{o}} \left\{ \mathbf{w}_i^T \mathbf{x} \,\middle\vert\, \max\limits_{j=1,m} \frac{\lvert \mathbf{Kx} - \hat{\mathbf{b}} \rvert_j}{s_j} \leqslant 1 \right\}$	$\hat{p}_i^{\,\mathrm{lo}} = \max\limits_{\mathbf{u}} \{ \mathbf{u}^T \mathbf{b} - \sum\limits_{j=1}^{m} \lvert u_j \rvert s_j \,\vert\, \mathbf{u}^T \mathbf{K} \leqslant \mathbf{w}_i^T \}$
	$\hat{p}_i^{\,\mathrm{up}} = \max\limits_{\mathbf{x} \geqslant \mathbf{o}} \left\{ \mathbf{w}_i^T \mathbf{x} \,\middle\vert\, \max\limits_{j=1,m} \frac{\lvert \mathbf{Kx} - \hat{\mathbf{b}} \rvert_j}{s_j} \leqslant 1 \right\}$	$\hat{p}_i^{\,\mathrm{up}} = \min\limits_{\mathbf{u}} \{ \mathbf{u}^T \mathbf{b} + \sum\limits_{j=1}^{m} \lvert \mathbf{u}_j \rvert s_j \,\vert\, \mathbf{u}^T \mathbf{K} \geqslant \mathbf{w}_i^T \}$

$$\mathbf{S} = \begin{bmatrix} s_1 & & & 0 \\ & s_2 & & \\ & & \ddots & \\ 0 & & & s_m \end{bmatrix}$$

Table 4.2

Error Distribution	Primal	Dual
I	$\hat{p}_i^{lo} = \min_{\mathbf{x} \geq \mathbf{o}} \{\mathbf{w}_i^T \mathbf{x} \mid \lVert \mathbf{K}\mathbf{x} - \hat{\mathbf{b}} \rVert_{1,\mathbf{s}} \leq 1\}$	$\hat{p}_i^{lo} = \max_{\mathbf{u}} \{\mathbf{u}^T \hat{\mathbf{b}} - \lVert \mathbf{u} \rVert_{\infty,\mathbf{s}'} \mid \mathbf{u}^T \mathbf{K} \leq \mathbf{w}_i^T\}$
II	$\hat{p}_i^{lo} = \min_{\mathbf{x} \geq \mathbf{o}} \{\mathbf{w}_i^T \mathbf{x} \mid \lVert \mathbf{K}\mathbf{x} - \hat{\mathbf{b}} \rVert_{2,\mathbf{s}} \leq 1\}$	$\hat{\mathbf{p}}_i^{lo} = \max_{\mathbf{u}} \{\mathbf{u}^T \hat{\mathbf{b}} - \lVert \mathbf{u} \rVert_{2,\mathbf{s}'} \mid \mathbf{u}^T \mathbf{K} \leq \mathbf{w}_i^T\}$
III	$\hat{p}_i^{lo} = \min_{\mathbf{x} \geq \mathbf{o}} \{\mathbf{w}_i^T \mathbf{x} \mid \lVert \mathbf{K}\mathbf{x} - \hat{\mathbf{b}} \rVert_{\infty,\mathbf{s}} \leq 1\}$	$\hat{p}_i^{lo} = \max_{\mathbf{u}} \{\mathbf{u}^T \hat{\mathbf{b}} - \lVert \mathbf{u} \rVert_{1,\mathbf{s}'} \mid \mathbf{u}^T \mathbf{K} \leq \mathbf{w}_i^T\}$

$$\mathbf{s} = \begin{bmatrix} 1/s_1 \\ 1/s_2 \\ \vdots \\ 1/s_m \end{bmatrix}, \qquad \mathbf{s}' = \begin{bmatrix} s_1 \\ s_2 \\ \vdots \\ s_m \end{bmatrix}$$

for a positive-definite matrix $\mathbf{S}^2$, and if we seek to estimate a linear function of the form

$$\bar{\phi} = \mathbf{w}^T \bar{\mathbf{x}}, \tag{4.4-40}$$

then the best linear unbiased estimate is given by

$$\hat{\phi}_{\text{blue}} = \hat{\mathbf{u}}^T \hat{\mathbf{b}}$$

where $\hat{\mathbf{u}}$ is the vector defined by

$$\hat{\mathbf{u}}^T \mathbf{S}^2 \hat{\mathbf{u}} = \min_{\mathbf{u}} \{\mathbf{u}^T \mathbf{S}\mathbf{u} | \mathbf{u}^T \mathbf{K} = \mathbf{w}^T\}.$$

This is the Markov formulation of the problem, and it corresponds to the dual formulation of the constrained linear estimation problem. The solution can also be written

$$\hat{\phi}_{\text{blue}} = \mathbf{w}^T \hat{\mathbf{x}}$$

where

$$\{\hat{\mathbf{x}}\} = \{\mathbf{x} | (\mathbf{K}\mathbf{x} - \hat{\mathbf{b}})^T \mathbf{S}^{-2} (\mathbf{K}\mathbf{x} - \hat{\mathbf{b}}) = \min = r_0\},$$

which is the Gauss formulation and corresponds to the primal problem. In the classical estimation problem, the matrix **K** has full rank, so that there is a single unique **x** for which the minimum value r_0 is assumed. Therefore, the set in the last expression contains only one vector $\hat{\mathbf{x}}$.

In Chapter 2 we also obtained confidence intervals of the form

$$I_\kappa(\bar{\phi}) = [\hat{\phi}_\kappa^{\text{lo}}, \hat{\phi}_\kappa^{\text{up}}] = [\hat{\phi}_{\text{blue}} - \kappa\sigma(\hat{\phi}_{\text{blue}}), \hat{\phi}_{\text{blue}} + \kappa\sigma(\hat{\phi}_{\text{blue}})]$$

where κ is a constant related to the probability that the interval actually contains the true value $\bar{\phi}$. We showed that corresponding to each value of κ there is a κ-ellipsoid in the x-space whose equation is

$$(\mathbf{K}\mathbf{x} - \hat{\mathbf{b}})^T \mathbf{S}^{-2} (\mathbf{K}\mathbf{x} - \hat{\mathbf{b}}) = r_0 + \kappa^2,$$

and that $\hat{\phi}_\kappa^{\text{lo}}$ and $\hat{\phi}_\kappa^{\text{up}}$ are just the two values assumed by the function $\mathbf{w}^T\mathbf{x}$ on the two support planes of the κ-ellipsoid that are orthogonal to the vector **w**. (See Fig. 4.16). An equivalent way of saying this is that $\hat{\phi}_\kappa^{\text{lo}}$ and $\hat{\phi}_\kappa^{\text{up}}$ are just the minimum and maximum values assumed by the function $\mathbf{w}^T\mathbf{x}$ when **x** is constrained to lie on the boundary of the κ-elipsoid. Thus we can write

$$\hat{\phi}_\kappa^{\text{lo}} = \min_{\mathbf{x}} \{\mathbf{w}^T \mathbf{x} | (\mathbf{K}\mathbf{x} - \hat{\mathbf{b}})^T \mathbf{S}^{-2} (\mathbf{K}\mathbf{x} - \hat{\mathbf{b}}) = r_0 + \kappa^2\} \tag{4.4-41}$$

and

$$\hat{\phi}_\kappa^{\text{up}} = \max_{\mathbf{x}} \{\mathbf{w}^T \mathbf{x} | (\mathbf{K}\mathbf{x} - \hat{\mathbf{b}})^T \mathbf{S}^{-2} (\mathbf{K}\mathbf{x} - \hat{\mathbf{b}}) = r_0 + \kappa^2\}. \tag{4.4-42}$$

These two equations are statements of the primal problems for determining the end points of the confidence interval $I_\kappa(\bar{\phi})$. They are analogous to Eqs. (4.4-12) and (4.4-32), which are the primal statements for the constrained linear estimation problem. The main differences between the two problems are that in the classical statistical problem there is no nonnegative constraint on the vector **x**, and that the constraint

$$(\mathbf{Kx} - \hat{\mathbf{b}})^T \mathbf{S}^{-2}(\mathbf{Kx} - \hat{\mathbf{b}}) = r_0 + \kappa^2$$

in the classical problem is replaced by the constraint

$$(\mathbf{Kx} - \hat{\mathbf{b}})^T \mathbf{S}^{-2}(\mathbf{Kx} - \hat{\mathbf{b}}) \leqslant 1$$

for the problems in the present chapter. The reason for this second difference is that in the classical problem we are dealing with a random variable **x** that can

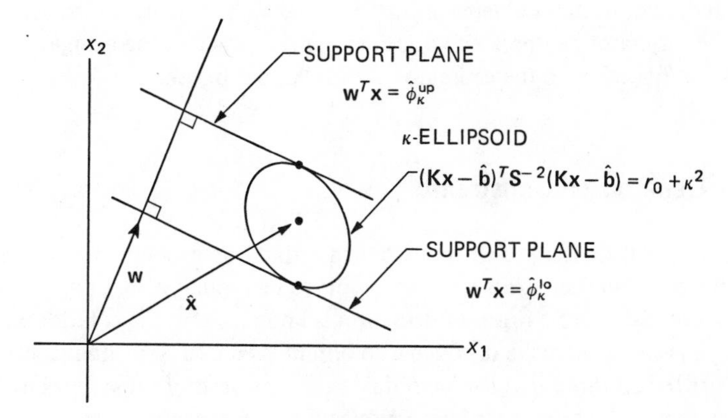

Figure 4.16.

take on *any* value with some nonzero probability, and the interval $I_\kappa(\bar{\phi})$ is a confidence interval that contains the value $\bar{\phi}$ with some probability less than 1, whereas in the present chapter we have been dealing with problems in which the vector **x** is guaranteed to lie in the constraint region and the interval $I_{\hat{p}_i}$ is guaranteed to contain the value $\bar{p}$.

If the Wolfe duality theorem is applied to the statement (4.4-41), the resulting dual problem will be

$$\hat{\phi}_\kappa^{lo} = \max_{\mathbf{u}} \{\mathbf{u}^T\hat{\mathbf{b}} - (r_0 + \kappa^2)^{1/2}(\mathbf{u}^T\mathbf{S}^2\mathbf{u})^{1/2} | \mathbf{u}^T\mathbf{K} = \mathbf{w}^T\}. \quad (4.4\text{-}43)$$

Similarly, the dual statement of (4.4-42) is

$$\hat{\phi}_\kappa^{up} = \min_{\mathbf{u}} \{\mathbf{u}^T\hat{\mathbf{b}} + (r_0 + \kappa^2)^{1/2}(\mathbf{u}^T\mathbf{S}^2\mathbf{u})^{1/2} | \mathbf{u}^T\mathbf{K} = \mathbf{w}^T\}. \quad (4.4\text{-}44)$$

These last two equations are the classical statistical analogues of Eqs. (4.4-31) and (4.4-34). The differences are the factor $(r_0 + \kappa^2)^{1/2}$, which appears in the objective functions of the statistical problems, and the replacement of the equality constraint

$$\mathbf{u}^T \mathbf{K} = \mathbf{w}^T \tag{4.4-45}$$

in the classical problems by the inequality constraints

$$\mathbf{u}^T \mathbf{K} \leqslant \mathbf{w}_i^T \tag{4.4-46}$$

and

$$\mathbf{u}^T \mathbf{K} \geqslant \mathbf{w}_i^T \tag{4.4-47}$$

in the constrained linear estimation problems. In the classical problem, the constraint (4.4-45) is an expression of the requirement that the estimator for $\bar{\phi}$ be unbiased. In the problems in this chapter, we have replaced the requirement that the estimator be unbiased with the requirements that the estimator for $\hat{p}_i^{\text{lo}}$ be "lower biased" and the estimator $\hat{p}_i^{\text{up}}$ be "upper biased."

4.5 Generalized Constraints

So far in our discussion of the constrained estimation problem, we have assumed that the solution vector $\bar{\mathbf{x}}$ is known a priori to be nonnegative. In many applications, even more restrictive a priori information is known about the solution $\bar{\mathbf{x}}$. These stronger constraints make it possible to obtain even narrower interval estimates $[\hat{p}_i^{\text{lo}}, \hat{p}_i^{\text{up}}]$ than those that can be obtained on the basis of the nonnegativity constraint alone. It is often possible to reduce these stronger constraints to a simple nonnegativity constraint by a linear transformation of variables. For example, suppose it is known a priori that the elements of the solution vector $\bar{\mathbf{x}}$ must be monotonically increasing as well as nonnegative. If we replace the variables x_1, $x_2, \ldots, x_n$ by new variables $q_1, q_2, \ldots, q_n$, which are defined by

$$\begin{aligned}
x_1 &= q_1,\\
x_2 &= x_1 + q_2,\\
x_3 &= x_2 + q_3,\\
&\cdots,\\
x_n &= x_{n-1} + q_n,
\end{aligned}$$

and if we require that

$$q_1 \geqslant 0, \quad q_2 \geqslant 0, \quad \ldots, \quad q_n \geqslant 0,$$

then the elements of the vector $\mathbf{x}$ must be nonnegative and monotonically nondecreasing. The vector $\mathbf{x}$ can be written

$$\mathbf{x} = \mathbf{Rq}, \qquad \mathbf{q} \geqslant \mathbf{0}, \tag{4.5-1}$$

where $\mathbf{R}$ is the $n \times n$ lower triangular matrix

$$\mathbf{R} = \begin{bmatrix} 1 & 0 & 0 & \dots & 0 & 0 \\ 1 & 1 & 0 & \dots & 0 & 0 \\ 1 & 1 & 1 & \dots & 0 & 0 \\ \vdots & \vdots & \vdots & & \vdots & \vdots \\ 1 & 1 & 1 & \dots & 1 & 0 \\ 1 & 1 & 1 & \dots & 1 & 1 \end{bmatrix}.$$

The end points of interval estimate $[\hat{p}_i^{\text{lo}}, \hat{p}_i^{\text{up}}]$ can be written, in terms of the new variables $\mathbf{q}$, as

$$\hat{p}_i^{\text{lo}} = \min_{\mathbf{q} \geqslant \mathbf{0}} \{\mathbf{w}_i^T \mathbf{Rq} \mid \|\mathbf{KRq} - \hat{\mathbf{b}}\|_{h,\mathbf{s}} \leqslant 1\}, \tag{4.5-2}$$

$$\hat{p}_i^{\text{up}} = \max_{\mathbf{q} \geqslant \mathbf{0}} \{\mathbf{w}_i^T \mathbf{Rq} \mid \|\mathbf{KRq} - \hat{\mathbf{b}}\|_{h,\mathbf{s}} \leqslant 1\}, \tag{4.5-3}$$

or as

$$\hat{p}_i^{\text{lo}} = \min_{\mathbf{q} \geqslant \mathbf{0}} \{(\mathbf{w}_i')^T \mathbf{q} \mid \|\mathbf{K}'\mathbf{q} - \hat{\mathbf{b}}\|_{h,\mathbf{s}} \leqslant 1\}, \tag{4.5-4}$$

$$\hat{p}_i^{\text{up}} = \max_{\mathbf{q} \geqslant \mathbf{0}} \{(\mathbf{w}_i')^T \mathbf{q} \mid \|\mathbf{K}'\mathbf{q} - \hat{\mathbf{b}}\|_{h,\mathbf{s}} \leqslant 1\} \tag{4.5-5}$$

where

$$(\mathbf{w}_i')^T = \mathbf{w}_i^T \mathbf{R} \tag{4.5-6}$$

and

$$\mathbf{K}' = \mathbf{KR}. \tag{4.5-7}$$

Thus the mathematical programming problem that must be solved has the same form as the problem for which there was only a nonnegativity constraint. The interval $[\hat{p}_i^{\text{lo}}, \hat{p}_i^{\text{up}}]$ that is obtained from problems (4.5-4) and (4.5-5) will be a subinterval of the interval that would be obtained using the nonnegativity constraint alone.

It is possible to formulate the constrained estimation problem for many other types of a priori constraints in the same form as (4.5-4) and (4.5-5), by using a linear transformation of the form (4.5-1). As another example, suppose it is known a priori that the $\bar{x}_i$ are monotonically increasing but not necessarily nonnegative. This can be accomplished by setting

$$\begin{aligned} x_1 \text{ (unconstrained)} &= q_1 - q_2, \\ x_2 &= x_1 + q_3, \\ x_3 &= x_2 + q_4, \\ &\cdots, \\ x_n &= x_{n-1} + q_{n+1} \end{aligned}$$

with $q_1 \geqslant 0, q_2 \geqslant 0, \ldots, q_n \geqslant 0, q_{n+1} \geqslant 0$. Then $\mathbf{x}$ is given by Eq. (4.5-1) and the constrained estimation problem can be written in the form of (4.5-4) and (4.5-5) if the matrix $\mathbf{R}$ is taken to be the $n \times (n+1)$ matrix

$$\mathbf{R} = \underbrace{\begin{bmatrix} 1 & -1 & 0 & 0 & \ldots & 0 & 0 \\ 1 & -1 & 1 & 0 & \ldots & 0 & 0 \\ 1 & -1 & 1 & 1 & \ldots & 0 & 0 \\ \vdots & \vdots & \vdots & \vdots & & \vdots & \vdots \\ 1 & -1 & 1 & 1 & \ldots & 1 & 0 \\ 1 & -1 & 1 & 1 & \ldots & 1 & 1 \end{bmatrix}}_{(n+1) \text{ columns}} \; n \text{ rows.}$$

As still another example, suppose it is known that the $\bar{x}_i$ are nonnegative and "smooth". Smoothness is a somewhat subjective quality that can be defined in many ways, but suppose that it is known that $\mathbf{x}$ must be at least as smooth as a positive combination of "triangular" vectors; that is,

$$\mathbf{x} = q_1 \begin{bmatrix} 1.0 \\ 0.75 \\ 0.50 \\ 0.25 \\ 0.0 \\ 0.0 \\ 0.0 \\ \vdots \\ 0.0 \\ 0.0 \end{bmatrix} + q_2 \begin{bmatrix} 0.75 \\ 1.0 \\ 0.75 \\ 0.50 \\ 0.25 \\ 0.0 \\ 0.0 \\ \vdots \\ 0.0 \\ 0.0 \end{bmatrix} + \ldots + q_5 \begin{bmatrix} 0.0 \\ 0.25 \\ 0.50 \\ 0.75 \\ 1.0 \\ 0.75 \\ 0.50 \\ \vdots \\ 0.0 \\ 0.0 \end{bmatrix} + \ldots + q_n \begin{bmatrix} 0.0 \\ 0.0 \\ 0.0 \\ 0.0 \\ 0.0 \\ 0.0 \\ 0.0 \\ \vdots \\ 0.75 \\ 1.0 \end{bmatrix}$$

Then, the matrix $\mathbf{R}$ is the $n \times n$ matrix

$$\mathbf{R}=\begin{bmatrix} 1.0 & 0.75 & 0.50 & 0.25 & 0.0 & \cdots & 0.0 & 0.0 \\ 0.75 & 1.0 & 0.75 & 0.50 & 0.25 & \cdots & 0.0 & 0.0 \\ 0.50 & 0.75 & 1.0 & 0.75 & 0.50 & \cdots & 0.0 & 0.0 \\ 0.25 & 0.50 & 0.75 & 1.0 & 0.75 & \cdots & 0.0 & 0.0 \\ 0.0 & 0.25 & 0.50 & 0.75 & 1.0 & \cdots & 0.0 & 0.0 \\ \vdots & \vdots & \vdots & \vdots & \vdots & & \vdots & \vdots \\ 0.0 & 0.0 & 0.0 & 0.0 & 0.0 & \cdots & 1.0 & 0.75 \\ 0.0 & 0.0 & 0.0 & 0.0 & 0.0 & \cdots & 0.75 & 1.0 \end{bmatrix}$$

and the constrained estimation problem is again given by Eqs. (4.5-4) and (4.5-5).

4.6 An Example of Constrained Estimation

In order to motivate a numerical example, we again consider the problem of Fox and Goodwin that was previously discussed in Section 1.5. The integral equation is

$$\int_0^1 (t^2+s^2)^{1/2}\, x(s)\, ds = \tfrac{1}{3}[(1+t^2)^{3/2}-t^3], \qquad 0 \leqslant t \leqslant 1.$$

This equation has the unique solution

$$x(s)=s, \qquad 0 \leqslant s \leqslant 1.$$

When the problem is discretized (using a Newton-Cotes-type quandrature) with mesh spacings

$$\Delta s = \frac{1-0}{n-1} = \frac{1}{n-1}, \qquad \Delta t = \frac{1-0}{m-1} = \frac{1}{m-1},$$

and mesh points

$$\begin{aligned} s_j &= 0+(j-1)\,\Delta s, \qquad & j&=1,2,\ldots,n, \\ t_i &= 0+(i-1)\,\Delta t, \qquad & i&=1,2,\ldots,m, \end{aligned}$$

the resulting linear system is

$$\mathbf{Kx} = \mathbf{y}$$

where $\mathbf{x}$ is the n-vector whose elements are the unknown quantities

$$x_j \cong x(s_j) = x\left(\frac{j-1}{n-1}\right), \qquad j = 1, 2, \ldots, n,$$

$\mathbf{y}$ is the m-vector of known quantities

$$y_i = y(t_i) = \tfrac{1}{3}[(1 + t_i^2)^{3/2} - t_i^3] = \left\{\left[1 + \left(\frac{i-1}{m-1}\right)^2\right]^{3/2} - \left(\frac{i-1}{m-1}\right)^3\right\},$$

$$i + 1, 2, \ldots, m,$$

and $\mathbf{K}$ is the $m \times n$ matrix whose elements are

$$K_{ij} = \alpha_j(t_i^2 + s_j^2)^{1/2}$$

$$= \alpha_j\left[\left(\frac{i-1}{m-1}\right)^2 + \left(\frac{j-1}{n-1}\right)^2\right]^{1/2},$$

$$i = 1, 2, \ldots, m, j = 1, 2, \ldots, n,$$

the α_j being the weighting coefficients for the particular quadrature formula used in the discretization. If we assume that the vector $\mathbf{y}$ is not known exactly, then the linear system can be written

$$\mathbf{Kx} = \hat{\mathbf{y}} - \hat{\mathbf{e}}$$

where $\hat{\mathbf{e}}$ is the error or uncertainty in the given vector $\hat{\mathbf{y}}$. If we further assume that the error vector is a random sample chosen from one of the three distributions I, II, or III [Eqs. (4.1-5), (4.1-6), or (4.1-7)], then we can apply the constrained estimation techniques described in the three preceding sections in order to determine upper and lower bounds for any linear function of the x_i. In particular, if we want lower and upper bounds for each of the x_i themselves, we can choose the window matrix to be the nth-order identity matrix; that is,

$$\mathbf{W} = \mathbf{I}_n$$

or

$$\begin{aligned}
\mathbf{w}_1^T &= \mathbf{e}_1^T = (1, 0, 0, \ldots, 0)\\
\mathbf{w}_2^T &= \mathbf{e}_2^T = (0, 1, 0, \ldots, 0)\\
\vdots\ & \quad\ \vdots \qquad \vdots\ \ \vdots\ \ \vdots\ \ \vdots\ \ \vdots\\
\mathbf{w}_n^T &= \mathbf{e}_n^T = (0, 0, 0, \ldots, 1),
\end{aligned}$$

and then apply the contrained estimation technique to each of the linear functions

$$x_j = p_j = \mathbf{w}_j^T \mathbf{x} = \mathbf{e}_j^T \mathbf{x}$$

in succession. The resulting interval $[x_j^{\text{lo}}, x_j^{\text{up}}]$, in each case, will be a guaranteed interval for the value of the x_j if we consider only the discrete linear system itself. But it should be noted that, because of the discretization error, the interval $[x_j^{\text{lo}}, x_j^{\text{up}}]$ is not a guaranteed interval for the true value $x(s_j)$ of the solution function of the integral equation. Of course, if enough mesh points are taken in the discretization, then the effect of the discretization error will be small compared to the effect of the uncertainty in $y(t)$ and the ill-conditioning of the problem, so that the resulting interval will be a good approximation to a guaranteed interval for $x(s_j)$. In Chapter 6, we will discuss a method for obtaining rigorous intervals for the solutions of Fredholm equations; but for the present, we will be content to obtain approximate intervals in order to illustrate the method of constrained estimation within a fairly simple framework.

In particular, we consider the case where the errors in the right-hand side are of type III; that is,

$$\max_{1 \leqslant i \leqslant m} \left\{ \frac{|e_i|}{s_i} \right\} \leqslant 1.$$

The constrained estimation problem in this case can be written

$$x_j^{\text{lo}} = \min_{\mathbf{x} \geqslant \mathbf{o}} \left\{ \mathbf{e}_j^T \mathbf{x} \,\middle|\, \max_{1 \leqslant j \leqslant m} \left[\frac{|\mathbf{K}\mathbf{x} - \hat{\mathbf{y}}|_j}{s_j} \right] \leqslant 1 \right\},$$

$$x_j^{\text{up}} = \max_{\mathbf{x} \geqslant \mathbf{o}} \left\{ \mathbf{e}_j^T \mathbf{x} \,\middle|\, \max_{1 \leqslant j \leqslant m} \left[\frac{|\mathbf{K}\mathbf{x} - \hat{\mathbf{y}}|_j}{s_j} \right] \leqslant 1 \right\}.$$

In Section 4.3, we showed that these two problems are linear programming problems, and that an equivalent way of stating them is

$$x_j^{\text{lo}} = \min_{\mathbf{x} \geqslant \mathbf{o}} \left\{ \mathbf{e}_j^T \mathbf{x} \,\middle|\, \begin{pmatrix} \mathbf{K} \\ -\mathbf{K} \end{pmatrix} \mathbf{x} \leqslant \begin{pmatrix} \hat{\mathbf{y}} + \mathbf{s}' \\ -\hat{\mathbf{y}} + \mathbf{s}' \end{pmatrix} \right\},$$

$$x_j^{\text{up}} = \max_{\mathbf{x} \geqslant \mathbf{o}} \left\{ \mathbf{e}_j^T \mathbf{x} \,\middle|\, \begin{pmatrix} \mathbf{K} \\ -\mathbf{K} \end{pmatrix} \mathbf{x} \leqslant \begin{pmatrix} \hat{\mathbf{y}} + \mathbf{s}' \\ -\hat{\mathbf{y}} + \mathbf{s}' \end{pmatrix} \right\},$$

where $\mathbf{s}'$ is the m-vector

$$\mathbf{s}' = (s_1, s_2, \ldots, s_m)^T.$$

It is a convenient mnemonic device to write the basic data of these two problems in tableau form as follows.

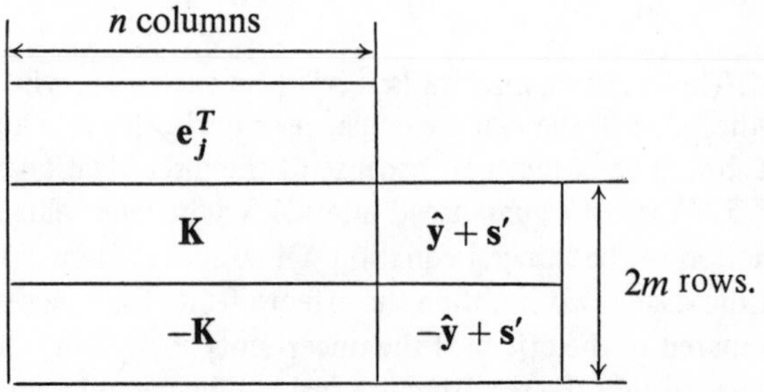

n columns		
$\mathbf{e}_j^T$		
$\mathbf{K}$	$\hat{\mathbf{y}}+\mathbf{s}'$	$2m$ rows.
$-\mathbf{K}$	$-\hat{\mathbf{y}}+\mathbf{s}'$	

In the next chapter we will discuss the methods actually used to calculate solutions of such linear programming problems. For the present, we are concerned only with the results of such calculations. All the results in this section were obtained by using a standard linear programming subroutine coded for a digital computer.

For the discretization, we chose mesh spacings with $m=5$ and $n=5$ (i.e., $\Delta t = 0.25 = \Delta s$), and we chose to use Simpson's rule as the quadrature formula. Thus the quadrature weights are

$$(\alpha_1, \alpha_2, \alpha_3, \alpha_4, \alpha_5) = \frac{\Delta s}{3}(1,4,2,4,1)$$
$$= (\tfrac{1}{12}, \tfrac{1}{3}, \tfrac{1}{6}, \tfrac{1}{3}, \tfrac{1}{12}).$$

The formulation of the problem demands not an exact knowledge of the right-hand side $y(t)$, but rather, for each value of i, an interval $[\hat{y}_i - s_i, \hat{y}_i + s_i]$ that is guaranteed to contain the true value of $y(t_i)$. For the purposes of this numerical example, the values $\hat{y}_i$ were chosen to be equal to the true values $y(t_i)$ and the errors s_i' were determined by using a random number table with the constraint that each error lie between 2.5% and 3.5% of the corresponding true value $y(t_i)$. The actual right-hand-side vector used in the linear programming problems was

$$\begin{bmatrix} \\ \hat{\mathbf{y}}+\mathbf{s}' \\ \\ \hline \\ -\hat{\mathbf{y}}+\mathbf{s}' \\ \\ \end{bmatrix} = \begin{bmatrix} 0.34369 \\ 0.37016 \\ 0.43895 \\ 0.52757 \\ 0.62764 \\ \hline -0.32298 \\ -0.34956 \\ -0.40942 \\ -0.49327 \\ -0.59131 \end{bmatrix}.$$

The accuracy used for all of the input numbers in this problem was five significant digits.

Thus far we have formulated the constrained estimation problem using only the simple nonnegativity constraint $\mathbf{x} \geqslant \mathbf{0}$. In the preceding section, we saw how to add other constraints by making a linear transformation of variables of the form

$$\mathbf{x} = \mathbf{R}\mathbf{q}, \qquad \mathbf{q} \geqslant \mathbf{0},$$

where $\mathbf{R}$ is a matrix designed to build in the new constraints. For the present problem, the tableau would then have the form

$\mathbf{e}_j^T \mathbf{R}$	
$\mathbf{KR}$	$\hat{\mathbf{y}} + \mathbf{s}'$
$-\mathbf{KR}$	$-\hat{\mathbf{y}} + \mathbf{s}'$

The matrix that produces a nonnegative and monotonic nondecreasing constraint is

$$\mathbf{R}_1 = \begin{bmatrix} 1 & 0 & 0 & 0 & 0 \\ 1 & 1 & 0 & 0 & 0 \\ 1 & 1 & 1 & 0 & 0 \\ 1 & 1 & 1 & 1 & 0 \\ 1 & 1 & 1 & 1 & 1 \end{bmatrix}$$

and the matrix that produces a nonnegative and smoothness constraint is

$$\mathbf{R}_2 = \begin{bmatrix} 1.00 & 0.75 & 0.50 & 0.25 & 0.00 \\ 0.75 & 1.00 & 0.75 & 0.50 & 0.25 \\ 0.50 & 0.75 & 1.00 & 0.75 & 0.50 \\ 0.25 & 0.50 & 0.75 & 1.00 & 0.75 \\ 0.00 & 0.25 & 0.50 & 0.75 & 1.00 \end{bmatrix}$$

Furthermore, the solution can be constrained to be nonnegative, monotonically nondecreasing, and smooth as well by taking $\mathbf{R}$ to be the product

$$\mathbf{R} = \mathbf{R}_1 \mathbf{R}_2.$$

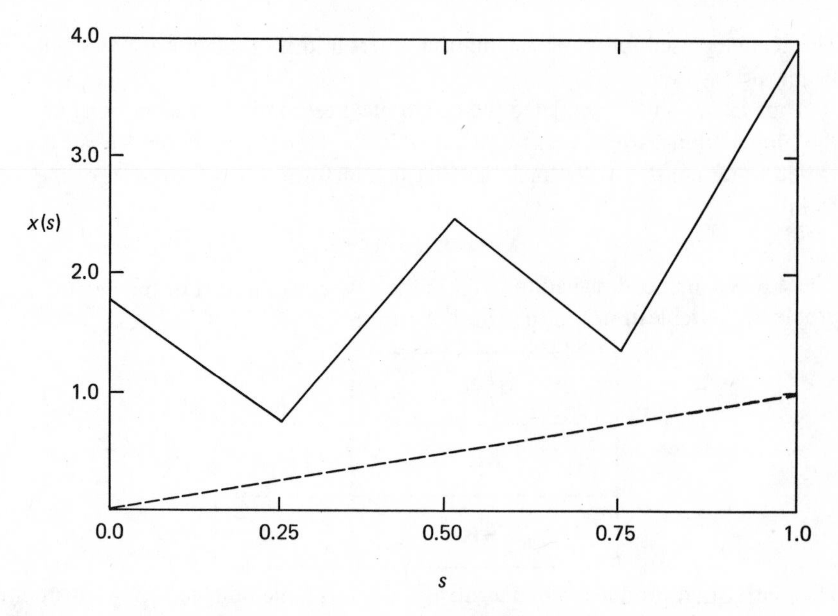

Figure 4.17. Upper and lower bounds to the solution of Fox–Goodwin problem constrained to be nonnegative. (Lower bound is zero line.)

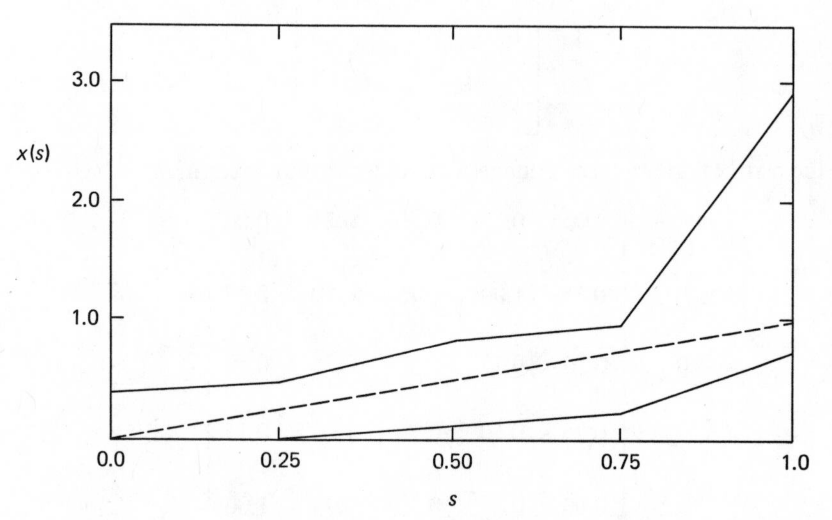

Figure 4.18. Upper and lower bounds to the solution of Fox–Goodwin problem constrained to be nonnegative and monotonically increasing.

The numerical results for the problem at hand are shown in Fig. 4.17 to 4.20. The graphs are all drawn to the same scale, and the individual points of the upper bounds and the lower bounds are joined by straight lines in order to aid visualization. The true solution $x(s)$ is drawn in as a dashed line. Figure 4.17 shows the result when only the nonnegativity constraint is used. The bounds obtained are certainly something less than very impressive. Somewhat better results (Fig. 4.18) are obtained when the monotonically nondecreasing constraint is added, but still the bounds obtained are not really very good. When the

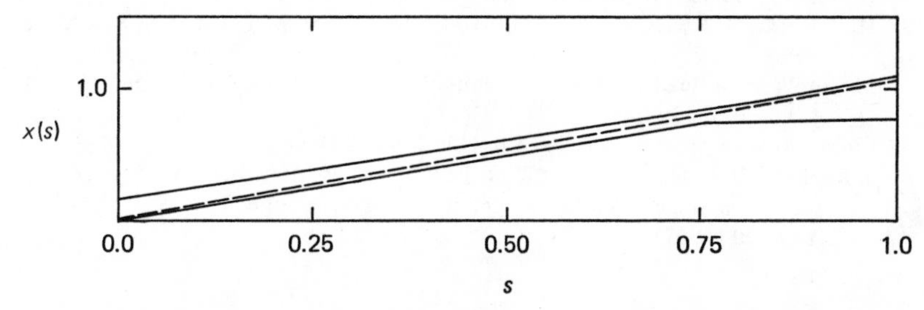

Figure 4.19. Upper and lower bounds to the solution of Fox-Goodwin problem constrained to be nonnegative and smooth.

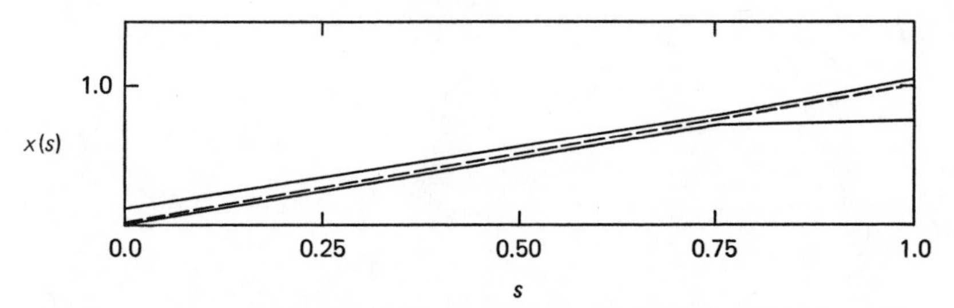

Figure 4.20. Upper and lower bounds to solution of Fox–Goodwin problem constrained to be nonnegative, smooth and monotonically increasing.

smoothness constraint is added to the nonnegativity constraint (Fig. 4.19), the results are significantly better and the bounds bracket the true solution much more closely than before. When the monotonically nondecreasing constraint is added to the nonnegative and smoothness constraints (Fig. 4.20), the resulting bounds do not represent a great deal of improvement over the case where only the nonnegativity and smoothness constraints were used, Thus, in the particular problem at hand, it appears that smoothness is a much more effective constraint for producing narrow bounds for the solution than is the monotonically nondecreasing constraint. Of course, this situation may well be reversed for some other problem.

The methods of this section have also been applied by one of the present authors to the integral equation of Bellman that was discussed in Section 1.5. The results can be found in a paper in the *Journal of Mathematical Analysis and Applications* [5].

References

1 Saul I. Gass, *Linear Programming,* McGraw-Hill, New York, 1958.
2 H. P. Kunzi, W. Krelle, and W. Oettli, *Nonlinear Programming,* Blaisdell, New York, 1966.
3 Phillip Wolfe, A duality theorem for non-linear programming, *Quart. Appl. Math.* **19** (3) (1961), 239–244.
4 Charles Schneeberger, Oak Ridge Nat. Lab. Internal Memorandum, July, 1966.
5 J. Replogle, B. D. Holcomb, and W. R. Burrus, The use of mathematical programming for solving singular and poorly conditioned systems of equations, *J. Math. Anal. Appl.* **20** (1967), 310–323.

CHAPTER 5

Mathematical Programming

5.1 The Linear Programming Problem and the Simplex Method

A linear programming problem is one in which it is required to optimize a linear function of several unknown variables that are subject to a set of linear constraints. One frequently occurring form of the problem can be written

$$\phi_0 = \max_{\mathbf{x} \geqslant \mathbf{0}} \{\mathbf{c}^T \mathbf{x} | \mathbf{A}\mathbf{x} \leqslant \mathbf{b}, \mathbf{b} \geqslant \mathbf{0}\} \tag{5.1-1}$$

where $\mathbf{c}$ is a given n-vector, $\mathbf{b}$ is a given nonnegative m-vector, and $\mathbf{A}$ is a given $m \times n$ matrix. The function

$$\phi(\mathbf{x}) = \mathbf{c}^T \mathbf{x} \tag{5.1-2}$$

is called the *objective function*, and the conditions

$$\mathbf{A}\mathbf{x} \leqslant \mathbf{b}, \tag{5.1-3}$$

$$\mathbf{x} \geqslant \mathbf{0} \tag{5.1-4}$$

are called the constraints on the problem. In other formulations, the problem is often a minimization rather than a maximization; and the inequality constraint (5.1-3) is often replaced by a constraint of the form

$$\mathbf{A}\mathbf{x} \geqslant \mathbf{b}$$

or by an equality constraint

$$\mathbf{A}\mathbf{x} = \mathbf{b}$$

or by some mixture of equalities and inequalities. The nonnegativity constraint (5.1-4) is a common feature of all linear programming problems. In this section, we will be chiefly concerned with the particular form (5.1-1) of the linear programming problem, but the other forms can all be treated in a completely similar manner.

The quantity ϕ_0 in (5.1-1) is just the maximum value that the objective function $\phi(\mathbf{x})$ can assume when $\mathbf{x}$ is required to lie in the region defined by the constraints. The inequality constraint (5.1-3) can be replaced with an equality by introducing an m-vector s defined by

$$\mathbf{A}\mathbf{x} + \mathbf{s} = \mathbf{b}. \tag{5.1-5}$$

The components $s_1, s_2, \ldots, s_m$ of s are called *slack variables* and are required to be nonnegative. Using the slack vector, we can rewrite the linear programming problem (5.1-1) in the form

$$\phi_0 = \max_{(\mathbf{x}^T,\mathbf{s}^T)\geqslant \mathbf{0}} \left\{ (\mathbf{c}^T, \mathbf{0}) \begin{pmatrix} \mathbf{x} \\ \mathbf{s} \end{pmatrix} \middle| (\mathbf{A}, \mathbf{I}_m) \begin{pmatrix} \mathbf{x} \\ \mathbf{s} \end{pmatrix} = \mathbf{b}, \mathbf{b} \geqslant \mathbf{0} \right\} \tag{5.1-6}$$

where $(\mathbf{c}^T, \mathbf{0})$ and $(\mathbf{x}^T, \mathbf{s}^T)$ are $(n+m)$-vectors and $(\mathbf{A}, \mathbf{I}_m)$ is an $m \times (n+m)$ composite matrix formed by adjoining an mth-order identity matrix to the matrix **A**.

Any vector $(\mathbf{x}^T, \mathbf{s}^T)$ that satisfies the constraints

$$\begin{pmatrix} \mathbf{x} \\ \mathbf{s} \end{pmatrix} \geqslant \mathbf{0}, \qquad \mathbf{A}\mathbf{x} + \mathbf{s} = \mathbf{b} \tag{5.1-7}$$

is said to be a *feasible solution* of the problem. A feasible solution is said to be a *basic feasible solution* if it has no more than m positive elements (the other elements having the value zero), and a basic feasible solution is said to be a *nondegenerate basic feasible solution* if it has exactly m positive (nonzero) elements. The solution that we seek is the *maximum feasible solution,* which is the feasible solution that maximizes the objective function

$$\phi(\mathbf{x}) = (\mathbf{c}^T, \mathbf{0}) \begin{pmatrix} \mathbf{x} \\ \mathbf{s} \end{pmatrix}. \tag{5.1-8}$$

It can be shown [1, Chapter 3] that the set of feasible solutions is a convex set K in $(n+m)$-space. This convex set will always be one of the following: (a) the null set; (b) a closed, bounded convex polygon; or (c) an open-ended convex polygon that is unbounded in some direction.

If K is the null set, then no solution to the problem exists. If K is a closed, bounded convex polygon, then it can be shown that a finite solution exists and further that the solution vector is the radius vector of one of the extreme points (or vertices) of K. If K is unbounded in some direction, the problem may or may not have a finite solution; if it does have a finite solution, the solution vector will be the radius vector to some vertex of K.

The three possibilities for K are illustrated in Figs. 5.1–5.3 for some hypothetical problems in which $m = 3$ and $n = 2$. Figure 5.1 illustrates the case where K is the null set. The three lines corresponding to the three linear equations

$$(\mathbf{A}\mathbf{x})_i = b_i, \qquad i = 1, 2, 3,$$

are shown with attached arrows pointing, in each case, into the half-space that satisfies the corresponding inequality constraint

$$(\mathbf{Ax})_i \leqslant b_i.$$

The shaded triangle is the intersection of these three half-spaces and hence is the region satisfying the constraint inequality (5.1-3). But this triangle does not have any points in common with the positive quadrant, which is the region defined by the constraint inequality (5.1-4). Thus K, which is the intersection of these two constraint regions, is the null set and hence no feasible solutions exist.

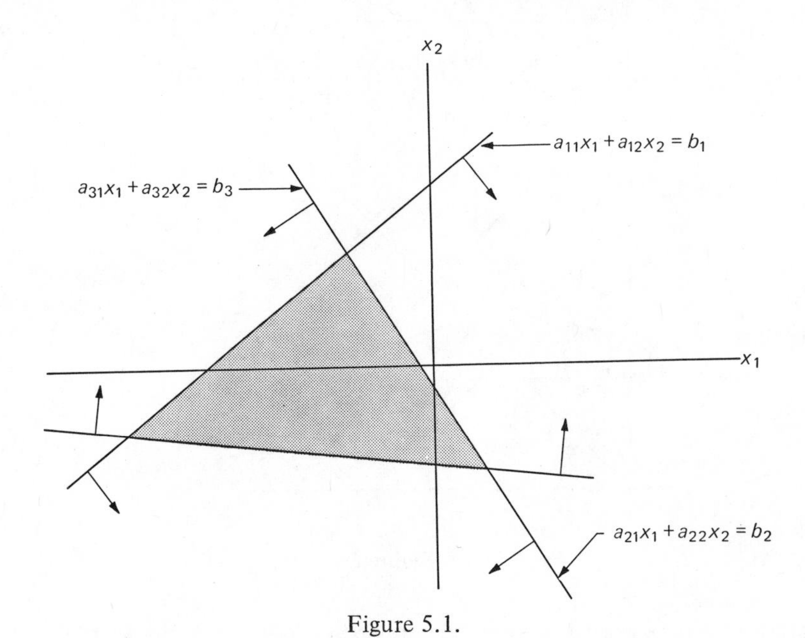

Figure 5.1.

Figure 5.2 illustrates the case where K is a closed convex polygon. In Fig. 5.2a, K is shown as the intersection of the region defined by the constraints

$$(\mathbf{Ax})_i \geqslant b_i, \qquad i = 1, 2, 3,$$

and the positive quadrant. Figure 5.2b shows the region K, the vector $\mathbf{c}$, and some of the level surfaces of the objective function $\phi(\mathbf{x})$. The level surfaces in this case are just straight lines orthogonal to the vector $\mathbf{c}$. In higher-dimensional problems, the level surfaces are $(n-1)$-dimensional hyperplanes orthogonal to $\mathbf{c}$. In the present case, $\mathbf{c}$ is a vector with all positive elements, so that the constant value of $\mathbf{c}^T\mathbf{x}$ that is associated with a given level surface increases as the distance of that level surface from the origin increases. The maximum value that can be

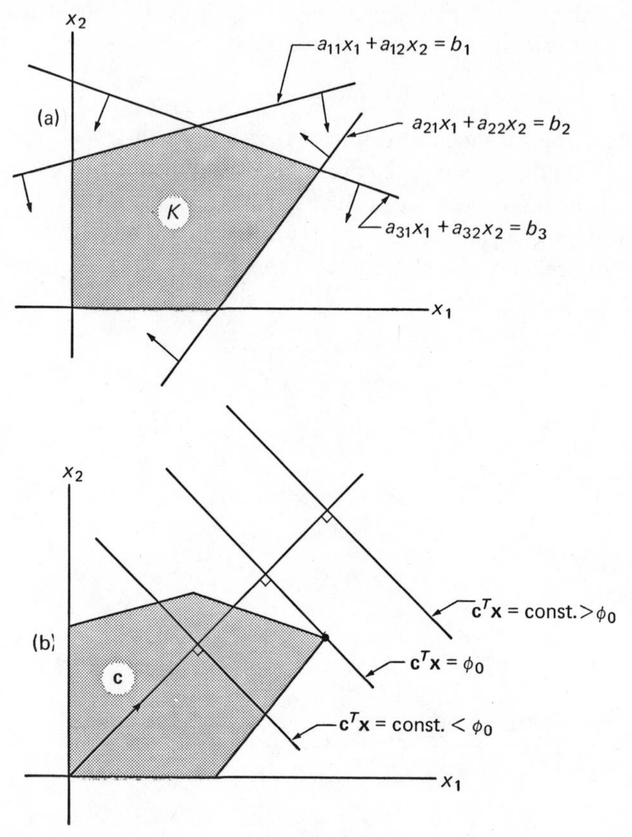

Figure 5.2

attained by $\mathbf{c}^T\mathbf{x}$ when $\mathbf{x}$ is constrained to be in K is just the value ϕ_0 that is associated with the level surface that passes through the vertex P.

Figure 5.3 illustrates the case where K is an open-ended, unbounded convex polygon. In this case, the objective function

$$\phi(\mathbf{x}) = \mathbf{c}^T \mathbf{x}$$

will have a maximum value in K (at the vertex P), but the objective function

$$\phi'(\mathbf{x}) = \mathbf{c}'^T \mathbf{x}$$

is unbounded in K.

The maximum value of the objective function $\mathbf{c}^T\mathbf{x}$ always occurs at a vertex of the convex set K. If the maximum is attained at more than one vertex, then the function will have this same value at every point that is a convex combination of these vertices. This situation is illustrated for a 3 x 2 example in Fig. 5.4. The

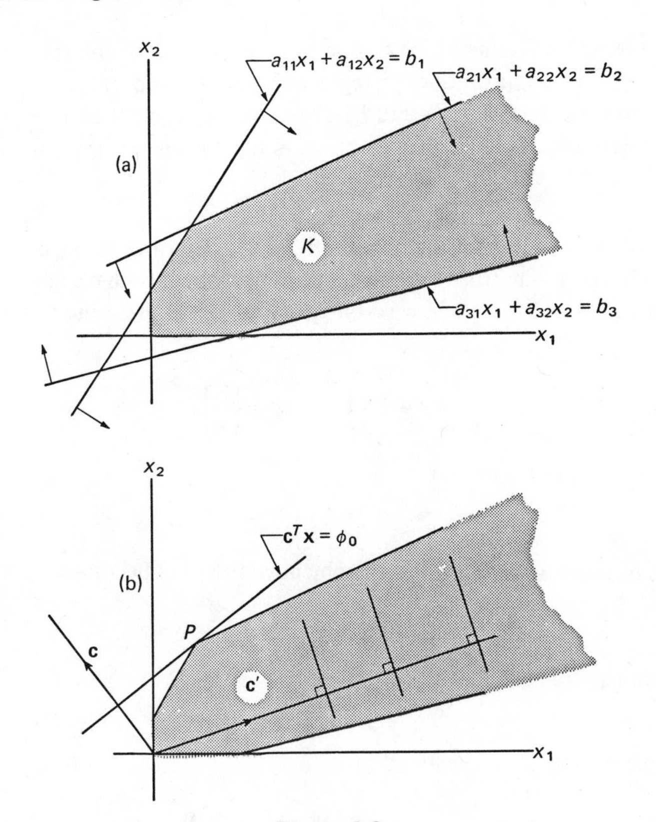

Figure 5.3.

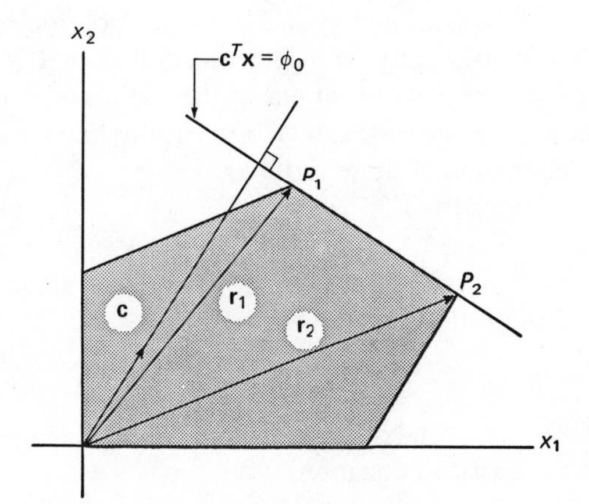

Figure 5.4.

objective function attains its maximum value ϕ_0 at both of the vertices P_1 and P_2 and also at every point on the line segment between them. If $\mathbf{r}_1$ and $\mathbf{r}_2$ are the radius vectors of the points P_1 and P_2, then the line segment between them is just the set of all possible convex combinations of $\mathbf{r}_1$ and $\mathbf{r}_2$, that is,

$$\{\mathbf{x}|\mathbf{x} = \gamma\mathbf{r}_1 + (1-\gamma)\mathbf{r}_2, 0 \leqslant \gamma \leqslant 1\},$$

and any vector $\mathbf{x}$ in this set is a maximum feasible solution of the problem.

So far in the geometrical exposition of the problem, we have ignored the slack variables $s_1, s_2, \ldots, s_m$. We now restate the problem by defining two new $(n+m)$-vectors

$$\mathbf{y} = \begin{pmatrix} \mathbf{x} \\ \mathbf{s} \end{pmatrix}, \qquad \mathbf{g} = \begin{pmatrix} \mathbf{c} \\ \mathbf{0} \end{pmatrix}, \tag{5.1-9}$$

and a new $m \times (n+m)$ matrix

$$\mathbf{P} = (\mathbf{A}, \mathbf{I}_m). \tag{5.1-10}$$

In terms of these new quantities, the objective function becomes

$$\phi(\mathbf{x}) = \phi(\mathbf{y}) = \mathbf{g}^T\mathbf{y} \tag{5.1-11}$$

and the constraints become

$$\mathbf{y} \geqslant \mathbf{0}, \qquad \mathbf{Py} = \mathbf{b}. \tag{5.1-12}$$

The problem itself can be restated as

$$\phi_0 = \max_{\mathbf{y} \geqslant \mathbf{o}} \{\mathbf{g}^T\mathbf{y} | \mathbf{Py} = \mathbf{b}\}. \tag{5.1-13}$$

The region in y-space that is defined by the constraint equations (5.1-12) is a convex set, which will be denoted by L. The convex polygon K is just the orthogonal projection of L into the n-dimensional x-space. It can be shown that the maximum value of $\phi(\mathbf{y})$ occurs at one of the vertices of L, and if the maximum is attained at more than one vertex, then $\phi(\mathbf{y})$ has this same maximum value at every convex combination of those vertices.

Now, consider the matrix $\mathbf{P}$;

$$\mathbf{P} = (\mathbf{p}_1, \mathbf{p}_2, \ldots, \mathbf{p}_{n+m}). \tag{5.1-14}$$

The columns of $\mathbf{P}$ comprise a set

$$P = \{\mathbf{p}_1, \mathbf{p}_2, \ldots, \mathbf{p}_{n+m}\} \tag{5.1-15}$$

of m-vectors. There are a total of $n+m$ vectors in P, so it is a linearly dependent set. The system of constraint equations

$$\mathbf{Py} = (\mathbf{p}_1, \mathbf{p}_2, \ldots, \mathbf{p}_{n+m})\mathbf{y} = \mathbf{b} \tag{5.1-16}$$

can also be written in the form

$$\sum_{i=1}^{n+m} y_i \mathbf{p}_i = \mathbf{b}. \tag{5.1-17}$$

Since the $\mathbf{p}_i$ are linearly dependent, there will be many sets of coefficients y_1, $y_2, \ldots, y_{n+m}$ that satisfy this equation. Of course, we are only interested in those solutions for which all the y_i are nonnegative.

Although P is a linearly dependent set, it will have many subsets of vectors that are linearly independent. From the definition (5.1-10), it is clear that the rows of $\mathbf{P}$ are linearly independent, and from this fact it follows that it is possible to pick from the set P at least one subset that contains a total of m linearly independent vectors; in fact, there will be many such subsets of P. Since $\mathbf{b}$ is an m-vector, it can be expressed as a linear combination of the vectors in any of these subsets, and in some cases it may be possible to express $\mathbf{b}$ as a linear combination of less than m of the columns of P. For example, suppose for the sake of definiteness that the first $k\,(k \leq m)$ columns of $\mathbf{P}$ are linearly independent and that the vector $\mathbf{b}$ can be expressed as

$$\mathbf{b} = y_1^0 \mathbf{p}_1 + y_2^0 \mathbf{p}_2 + \ldots + y_k^0 \mathbf{p}_k$$

for some set of coefficients $y_1^0, y_2^0, \ldots, y_k^0$. In such cases, the m-vector $\mathbf{y}^0$ defined by

$$(\mathbf{y}^0)^T = (y_1^0, y_2^0, \ldots, y_k^0, \underbrace{0, \ldots, 0}_{(n+m-k)\ \text{zeros}})$$

is a solution to the system (5.1-16). If it is also true that $y_1^0, y_2^0, \ldots, y_k^0$ are all positive, then $\mathbf{y}^0$ is a *basic feasible solution* to the linear programming problem (5.1-13), and furthermore it can be shown [1, pp. 41, 42] that the point $\mathbf{y}^0$ is a vertex of the convex set L. More generally, if $\{\mathbf{p}_{i_1}, \mathbf{p}_{i_2}, \ldots, \mathbf{p}_{i_k}\}$ is any linearly independent subset of P for which it is possible to find positive coefficients $y_{i_1}^0, y_{i_2}^0, \ldots, y_{i_k}^0$ that satisfy

$$y_{i_1}^0 \mathbf{p}_{i_1}^0 + y_{i_2}^0 \mathbf{p}_{i_2} + \ldots + y_{i_k}^0 \mathbf{p}_{i_k} = \mathbf{b},$$

then the basic feasible solution vector $\mathbf{y}$, which is defined by

$$y_i = \begin{cases} y_i^0, & i = i_1, i_2, \ldots, i_k, \\ 0, & \text{otherwise,} \end{cases}$$

is the radius vector of some vertex of the convex set L.

This last result associates a vertex of L with every basic feasible solution vector. It can also be shown [1, pp. 42, 43] that a basic feasible solution can be associated with every vertex of L. If

$$\mathbf{y}^T = (y_1, y_2, \ldots, y_{n+m})$$

is the radius vector of a vertex of L, then at most m of the y_i are positive (the others are zero) and the vectors $\mathbf{p}_i$ that are associated with the positive y_i form a linearly independent subset of P. In fact, for any vertex of L, it is possible to pick from P an associated subset that contains exactly m linearly independent vectors, although in some cases the radius vector for the vertex may have less than m nonzero components. For each vertex, the associated set of m linearly independent columns of P is called the *basis set* because m linearly independent vectors form a basis for b-space. The nonzero components of the basic feasible solution vector corresponding to that vertex are called *basic variables* and the zero components are called *nonbasic variables.* If one or more of the basic variables in the basic feasible solution corresponding to the vertex is equal to zero, then the solution is said to be *degenerate.*

Since the maximum feasible solution always occurs at a vertex of L, one way to solve the linear programming problem would be to compute the value of the objective function for every vertex and then pick the vertex that gives the maximum value. This is not a very good way to solve the problem, since the number of vertices of the set L can be as high as

$$\binom{n+m}{m} = \frac{(n+m)!}{m!n!},$$

which is an extremely large number even for small values of m and n (e.g., for $m = n = 10$, the number above is 184,756). In order to avoid an excessive amount of computation, the usual strategy for solving the problem is to begin with some basic feasible solution and to proceed iteratively along the edges of L from vertex to vertex in such a way that each new vertex has a larger associated value of the objective function than the previous one. The iteration is continued until it is no longer possible to find a new vertex with a larger value of the objective function.

The most widely used iteration schemes are variants of the *simplex method,* which was originally developed by G. B. Dantzig in 1947. Suppose that by some means we have already obtained a basic feasible solution $\mathbf{y}^{(0)}$

$$(\mathbf{y}^{(0)})^T = (y_1^{(0)}, y_2^{(0)}, \ldots, y_m^{(0)}, \underbrace{0, 0, \ldots, 0}_{n \text{ zeros}}) \tag{5.1-18}$$

that has as its associated basis set the first m columns of $\mathbf{P}$, that is, $\{\mathbf{p}_1, \mathbf{p}_2, \ldots, \mathbf{p}_m\}$. Note that there is no loss of generality in assuming that the first m columns of $\mathbf{P}$ are linearly independent and correspond to a nondegenerate basic feasible solution with the form of $\mathbf{y}^{(0)}$ because, if we can find any nondegenerate basic feasible solution vector, then we can permute its elements until all the nonzero elements come first, and if we carry out the same permutations on the elements of the vector $\mathbf{g}$ and on the columns of the matrix $\mathbf{P}$, the result is a new problem that is completely equivalent to the old one and that has a solution vector and a basis set like the one we are assuming.

Each step of the simplex method will generate a new basis set by dropping one of the vectors currently in the basis and adding in its place one of the columns of **P** that is not currently in the basis. At each step ν of the iteration, we define a matrix $\mathbf{M}^{(\nu)}$ to be an $m \times m$ matrix whose columns are just the basis vectors for that step. Thus, since the initial basis set was just the first m columns of **P**, we take

$$\mathbf{M}^{(0)} = (\mathbf{p}_1, \mathbf{p}_2, \ldots, \mathbf{p}_m). \tag{5.1-19}$$

Similarly, the other matrices in the sequence $\mathbf{M}^{(1)}, \mathbf{M}^{(2)}, \mathbf{M}^{(3)}, \ldots$ can be obtained from **P** by simply dropping some combination of n columns and perhaps rearranging the remaining m columns. Of course, the remaining m columns must always be linearly independent and, furthermore, have the property that, when the vector **b** is written as a linear combination of them, the coefficient of each column must be nonnegative. Another way to say this is that the columns of $\mathbf{M}^{(\nu)}$ must be chosen from the columns of **P** in such a way that $\mathbf{M}^{(\nu)}$ is nonsingular, and the solution vector $\mathbf{y}_m^{(\nu)}$ of the system

$$\mathbf{M}^{(\nu)}\mathbf{y}_m^{(\nu)} = \mathbf{b}$$

does not have any negative elements. The reason for this last requirement is to ensure that the elements of $\mathbf{y}_m^{(\nu)}$ can be used as the nonzero elements of a basic feasible solution vector $\mathbf{y}^{(\nu)}$.

The actual calculations in the simplex method do not explicitly form the matrices $\mathbf{M}^{(\nu)}$, but rather operate at each step upon a matrix $\mathbf{T}^{(\nu)}$, which is called a *tableau.* The tableau for the νth step can be defined by

$$\mathbf{T}^{(\nu)} = (\mathbf{M}^{(\nu)})^{-1}\mathbf{P}. \tag{5.1-20}$$

Thus the initial tableau has the form

$$\mathbf{T}^{(0)} = (\mathbf{M}^{(0)})^{-1}\mathbf{P} = \begin{bmatrix} 1 & 0 & \ldots & 0 & t_{1,m+1}^{(0)} & \cdots & t_{1,m+n}^{(0)} \\ 0 & 1 & \ldots & 0 & t_{2,m+1}^{(0)} & \cdots & t_{2,m+n}^{(0)} \\ \vdots & \vdots & & \vdots & \vdots & & \vdots \\ 0 & 0 & \ldots & 1 & t_{m,m+1}^{(0)} & \cdots & t_{m,m+n}^{(0)} \end{bmatrix} \tag{5.1-21}$$

and every succeeding tableau will have m columns that are just the columns of the mth-order identity matrix, although they will not in general be adjacent to one another or in the same order as they appear in the identity matrix. Although the tableaux are defined by (5.1-20) in terms of the matrices $(\mathbf{M}^{(\nu)})^{-1}$, they are not calculated by this formula. Once the initial tableau $\mathbf{T}^{(0)}$ has been obtained, the simplex procedure obtains each new tableau from the preceding one by an elimination calculation that will be described below. Note that it is not really necessary to do any extended calculations to obtain $\mathbf{T}^{(0)}$, for if we rewrite the original problem (5.1-6) in the form

$$\phi_0 = \max_{(\mathbf{s}^T, \mathbf{x}^T) \geqslant \mathbf{o}} \left\{ (0, \mathbf{c}^T) \begin{pmatrix} \mathbf{s} \\ \mathbf{x} \end{pmatrix} \middle| (\mathbf{I}_m, \mathbf{A}) \begin{pmatrix} \mathbf{s} \\ \mathbf{x} \end{pmatrix} = \mathbf{b} \right\}$$

and take

$$\mathbf{P}' = (\mathbf{I}_m, \mathbf{A}), \qquad \mathbf{y}' = \begin{pmatrix} \mathbf{s} \\ \mathbf{x} \end{pmatrix}, \qquad \mathbf{g}' = \begin{pmatrix} \mathbf{0} \\ \mathbf{c} \end{pmatrix},$$

rather than

$$\mathbf{P} = (\mathbf{A}, \mathbf{I}_m), \qquad \mathbf{y} = \begin{pmatrix} \mathbf{x} \\ \mathbf{s} \end{pmatrix}, \qquad \mathbf{g} = \begin{pmatrix} \mathbf{c} \\ \mathbf{0} \end{pmatrix},$$

then the vector

$$\mathbf{y}'^{(0)} = \begin{pmatrix} \mathbf{s}^{(0)} \\ \mathbf{x}^{(0)} \end{pmatrix} = \begin{pmatrix} \mathbf{b} \\ \mathbf{0} \end{pmatrix}$$

is the required initial basic feasible solution and the matrix $\mathbf{M}'^{(0)}$ is just

$$\mathbf{M}'^{(0)} = \mathbf{I}_m.$$

Thus, the initial tableau is just

$$\mathbf{T}^{(0)} = (\mathbf{I}_m, \mathbf{A}) = \mathbf{P}'.$$

Although we have just seen that it is possible to obtain the initial tableau $\mathbf{T}^{(0)}$ very simply just by rewriting the problem, it is instructive to consider another method for obtaining it. We assume that the first m columns of $\mathbf{P}$ are linearly independent and, furthermore, that they correspond to a basic feasible solution $\mathbf{y}$. We also assume that they do not already form an identity matrix, and we calculate the required tableau by a *Gauss–Jordan reduction.* This process is a series of elementary row operations that are designed to reduce the leftmost $m \times m$ submatrix of $\mathbf{P}$ to an identity matrix. To illustrate the process, let the initial unreduced matrix $\mathbf{P}$ be denoted by

$$\mathbf{P} = \mathbf{P}^{(0)} = \begin{bmatrix} p_{11}^{(0)} & p_{12}^{(0)} & \cdots & p_{1m}^{(0)} & p_{1,m+1}^{(0)} & \cdots & p_{1,m+n}^{(0)} \\ p_{21}^{(0)} & p_{22}^{(0)} & \cdots & p_{2m}^{(0)} & p_{2,m+1}^{(0)} & \cdots & p_{2,m+n}^{(0)} \\ \vdots & \vdots & & \vdots & \vdots & & \vdots \\ p_{m1}^{(0)} & p_{m2}^{(0)} & \cdots & p_{mm}^{(0)} & p_{m,m+1}^{(0)} & \cdots & p_{m,m+n}^{(0)} \end{bmatrix}.$$

When the first column is reduced, the result is a matrix of the form

$$\mathbf{P}^{(1)}=\begin{bmatrix} 1 & p_{12}^{(1)} & \cdots & p_{1m}^{(1)} & p_{1,m+1}^{(1)} & \cdots & p_{1,m+n}^{(1)} \\ 0 & p_{22}^{(1)} & \cdots & p_{2m}^{(1)} & p_{2,m+1}^{(1)} & \cdots & p_{2,m+n}^{(1)} \\ \vdots & \vdots & & \vdots & \vdots & & \vdots \\ 0 & p_{m2}^{(1)} & \cdots & p_{mm}^{(1)} & p_{m,m+1}^{(1)} & \cdots & p_{m,m+n}^{(1)} \end{bmatrix}.$$

The reduction is accomplished by applying the following m elementary row operations.

$$i=1:\ p_{1j}^{(1)}=\frac{p_{1j}^{(0)}}{p_{11}^{(0)}},\qquad j=1,2,\ldots,m+n,$$

$$i=2,3,\ldots,m:\ p_{ij}^{(1)}=p_{ij}^{(0)}-\frac{p_{i1}^{(0)}}{p_{11}^{(0)}}p_{1j}^{(0)},\qquad j=1,2,\ldots,m+n. \tag{5.1-22}$$

The element $p_{11}^{(0)}$ that is used as a divisor in the foregoing expressions is called a *pivot.* The second step of the reduction is just like the first except that the pivot element is $p_{22}^{(1)}$. The row transformations for the second step are

$$i=2:\ p_{2j}^{(2)}=\frac{p_{2j}^{(1)}}{p_{22}^{(1)}},\qquad j=1,2,\ldots,m+n;$$

$$i=1,i=3,4,\ldots,m:\ p_{ij}^{(2)}=p_{ij}^{(1)}-\frac{p_{i2}^{(1)}}{p_{22}^{(1)}}p_{2j}^{(1)},\qquad j=1,2,\ldots,m+n, \tag{5.1-23}$$

and the result of the second step is

$$\mathbf{P}^{(2)}=\begin{bmatrix} 1 & 0 & p_{13}^{(2)} & \cdots & p_{1m}^{(2)} & p_{1,m+1}^{(2)} & \cdots & p_{1,m+n}^{(2)} \\ 0 & 1 & p_{23}^{(2)} & \cdots & p_{2m}^{(2)} & p_{2,m+1}^{(2)} & \cdots & p_{2,m+n}^{(2)} \\ 0 & 0 & p_{33}^{(2)} & \cdots & p_{3m}^{(2)} & p_{3,m+1}^{(2)} & \cdots & p_{3,m+n}^{(2)} \\ \vdots & \vdots & \vdots & & \vdots & \vdots & & \vdots \\ 0 & 0 & p_{m3}^{(2)} & \cdots & p_{mm}^{(2)} & p_{m,m+1}^{(2)} & \cdots & p_{m,m+n}^{(2)} \end{bmatrix}.$$

The reduction proceeds in this manner, from column to column, using at the kth step the pivot element $p_{kk}^{(k)}$ and the row transformations

$$i=k:\quad p_{kj}^{(k)}=\frac{p_{kj}^{(k-1)}}{p_{kk}^{(k-1)}},\qquad j=1,2,\ldots,m+n;$$

$$\begin{aligned}&i=1,2,\ldots,k-1,\\ &i=k+1,k+2,\ldots,m:\quad p_{ij}^{(k)}=p_{ij}^{(k-1)}-\frac{p_{ik}^{(k-1)}}{p_{kk}^{(k-1)}}\,p_{kj}^{(k-1)},\\ &\qquad\qquad j=1,2,\ldots,m+n.\end{aligned}\tag{5.1-24}$$

The result of m such steps is the desired tableau; that is,

$$\mathbf{P}^{(n)}=\begin{bmatrix}1 & 0 & \cdots & 0 & p_{1,m+1}^{(m)} & \cdots & p_{1,m+n}^{(m)}\\ 0 & 1 & \cdots & 0 & p_{2,m+1}^{(m)} & \cdots & p_{2,m+n}^{(m)}\\ \vdots & \vdots & & \vdots & \vdots & & \vdots\\ 0 & 0 & \cdots & 1 & p_{m,m+1}^{(m)} & \cdots & p_{m,m+n}^{(m)}\end{bmatrix}$$

$$=\begin{bmatrix}1 & 0 & \cdots & 0 & t_{1,m+1}^{(0)} & \cdots & t_{1,m+n}^{(0)}\\ 0 & 1 & \cdots & 0 & t_{2,m+1}^{(0)} & \cdots & t_{2,m+n}^{(0)}\\ \vdots & \vdots & & \vdots & \vdots & & \vdots\\ 0 & 0 & \cdots & 1 & t_{m,m+1}^{(0)} & \cdots & t_{m,m+n}^{(0)}\end{bmatrix}\tag{5.1-25}$$

The tableau in Eq. (5.1-25) is exactly the same as the one defined by Eq. (5.1-21). Carrying out the elementary row transformations described by Eqs. (5.1-22)-(5.1-24) is equivalent to forming the matrix $\mathbf{M}^{(0)}$ and carrying out the matrix inversion and multiplication indicated by Eq. (5.1-21).

The tableau $\mathbf{T}^{(\nu)}$ for each step ν of the simplex iteration can be computed in a similar manner without actually forming a new matrix $\mathbf{M}^{(\nu)}$, inverting it, and calculating the product matrix on the right-hand side of Eq. (5.1-20). To illustrate the method, let us assume, for the sake of definiteness, that for the first iteration we want to drop the vector $\mathbf{p}_2$ from the basis and replace it with the vector $\mathbf{p}_{m+1}$. Thus, we are assuming that the set of vectors

$$\{\mathbf{p}_1, \mathbf{p}_{m+1}, \mathbf{p}_3, \ldots, \mathbf{p}_m\}$$

is a linearly independent set corresponding to a basic feasible solution $\mathbf{y}^{(1)}$. The matrix $\mathbf{M}^{(1)}$ would then be

$$M^{(1)} = (\mathbf{p}_1, \mathbf{p}_{m+1}, \mathbf{p}_3, \ldots, \mathbf{p}_m),$$

and the new tableau would be

$$\mathbf{T}^{(1)} = (\mathbf{M}^{(1)})^{-1}\mathbf{P} = \begin{bmatrix} 1 & t_{12}^{(1)} & 0 & \cdots & 0 & 0 & t_{1,m+2}^{(1)} & \cdots & t_{1,m+n}^{(1)} \\ 0 & t_{22}^{(1)} & 0 & \cdots & 0 & 1 & t_{2,m+2}^{(1)} & \cdots & t_{2,m+n}^{(1)} \\ 0 & t_{32}^{(1)} & 1 & \cdots & 0 & 0 & t_{3,m+2}^{(1)} & \cdots & t_{3,m+n}^{(1)} \\ \vdots & \vdots & \vdots & & \vdots & \vdots & \vdots & & \vdots \\ 0 & t_{m2}^{(1)} & 0 & \cdots & 1 & 0 & t_{m,m+2}^{(1)} & \cdots & t_{m,m+n}^{(1)} \end{bmatrix}.$$

Exactly the same tableau would be obtained from $\mathbf{T}^{(0)}$ by using the element $t_{2,m+1}^{(0)}$ as a pivot and applying the elementary row transformations

$$i=2: \quad t_{2j}^{(1)} = \frac{t_{2j}^{(0)}}{t_{2,m+1}^{(0)}}, \qquad j=1,2,\ldots,m+n;$$

$$i=1, i=3,4,\ldots,m+n: \quad t_{ij}^{(1)} = t_{ij}^{(0)} - \frac{t_{i,m+1}^{(0)}}{t_{2,m+1}^{(0)}} t_{2,j}^{(0)}, \qquad j=1,2,\ldots,m+n.$$

Each succeeding iteration of the method uses a similar set of elementary row transformations, destroying one of the identity matrix columns but producing it again in some other column of the tableau. If the element $t_{r,s}^{(\nu-1)}$ is chosen as the pivot for the νth iteration, then the elementary row transformations are

$$i=r: \quad t_{r,j}^{(\nu)} = \frac{t_{r,j}^{(\nu-1)}}{t_{r,s}^{(\nu-1)}}, \qquad j=1,2,\ldots,m+n;$$

$$\begin{aligned} &i=1,2,\ldots,r-1, \\ &i=r+1,r+2,\ldots,m: \quad t_{i,j}^{(\nu)} = t_{i,j}^{(\nu-1)} - \frac{t_{i,s}^{(\nu-1)}}{t_{r,s}^{(\nu-1)}} t_{r,j}^{(\nu-1)}, \\ &\qquad\qquad j=1,2,\ldots,m+n. \end{aligned} \tag{5.1-26}$$

At each step, the pivot is chosen in such a way that the set of vectors that would comprise the columns of $\mathbf{M}^{(\nu)}$, if it were formed, is a basis set corresponding to a basic feasible solution.

The basic feasible solution vector for each step ν of the process will be denoted by $\mathbf{y}^{(\nu)}$. This vector must be nonnegative and satisfy the constraint equation

$$\mathbf{P}\mathbf{y}^{(\nu)} = \mathbf{b}.$$

Multiplying on both sides by $(\mathbf{M}^{(\nu)})^{-1}$gives

$$(\mathbf{M}^{(\nu)})^{-1}\,\mathbf{P}\mathbf{y}^{(\nu)} = (\mathbf{M}^{(\nu)})^{-1}\,\mathbf{b}.$$

Thus, by Eq. (5.1-20), the solution vector at each step must satisfy

$$\mathbf{T}^{(\nu)}\,\mathbf{y}^{(\nu)} = \mathbf{b}^{(\nu)} \tag{5.1-27}$$

where

$$\mathbf{b}^{(\nu)} = (\mathbf{M}^{(\nu)})^{-1}\,\mathbf{b}. \tag{5.1-28}$$

Substituting Eqs. (5.1-18) and (5.1-21) into Eq. (5.1-27) gives, for $\nu = 0$,

$$\begin{bmatrix} 1 & 0 & \cdots & 0 & t^{(0)}_{1,m+1} & \cdots & t^{(0)}_{1,m+n} \\ 0 & 1 & \cdots & 0 & t^{(0)}_{2,m+1} & \cdots & t^{(0)}_{2,m+n} \\ \vdots & \vdots & & \vdots & \vdots & \vdots & \vdots \\ 0 & 0 & \cdots & 1 & t^{(0)}_{m,m+1} & \cdots & t^{(0)}_{m,m+n} \end{bmatrix} \begin{bmatrix} y^{(0)}_1 \\ y^{(0)}_2 \\ \vdots \\ y^{(0)}_m \\ 0 \\ \vdots \\ 0 \end{bmatrix} = \begin{bmatrix} b^{(0)}_1 \\ b^{(0)}_2 \\ \vdots \\ b^{(0)}_m \end{bmatrix},$$

from which it follows that

$$y^{(0)}_i = b^{(0)}_i, \qquad i = 1, 2, \ldots, m. \tag{5.1-29}$$

The vector $\mathbf{b}^{(0)}$ is defined by Eq. (5.1-28) with $\nu = 0$, but it can also be obtained by carrying out the same elementary row operations on the vector $\mathbf{b}$ that were performed on the matrix $\mathbf{P}$ in order to obtain the initial tableau $\mathbf{T}^{(0)}$. Let the vector $\mathbf{b}$ be denoted by

$$\mathbf{b} = \mathbf{b}^{[0]}$$

and let $\mathbf{b}^{[1]}, \mathbf{b}^{[2]}, \ldots, \mathbf{b}^{[m]}$ denote the vectors obtained at each step of the Gauss–Jordan reduction when the elementary row transformations are extended to operate on the vector $\mathbf{b}$. From (5.1-22) it is clear that the vector $\mathbf{b}^{[1]}$ is obtained from $\mathbf{b}^{[0]}$ by

$$i=1: \quad b_1^{[1]} = \frac{b_1^{[0]}}{p_{11}^{(0)}};$$

$$i=2, 3, \ldots, m: \quad b_i^{[1]} = b_i^{[0]} - \frac{p_{i1}^{(0)}}{p_{11}^{(0)}} b_1^{[0]},$$

and from (5.1-24) it follows that in general $\mathbf{b}^{[k]}$ is obtained from $\mathbf{b}^{[k-1]}$ by

$$i=k: \quad b_k^{[k]} = \frac{b_k^{[k-1]}}{p_{kk}^{(k-1)}};$$

$$i=1, 2, \ldots, k-1,$$

$$i=k+1, k+2, \ldots, m: \quad b_i^{[k]} = b_i^{[k-1]} - \frac{p_{ik}^{(k-1)}}{p_{kk}^{(k-1)}} b_k^{[k-1]}. \qquad (5.1\text{-}30)$$

The result of m such steps is the required vector

$$\mathbf{b}^{(0)} = \mathbf{b}^{[m]}.$$

In a similar manner, the vector $\mathbf{b}^{(\nu)}$ for each step ν of the simplex iteration can be computed from the previous vector $\mathbf{b}^{(\nu-1)}$ by performing on $\mathbf{b}^{(\nu-1)}$ the same elementary row operations that are used to obtain $\mathbf{T}^{(\nu)}$ from $\mathbf{T}^{(\nu-1)}$. If the pivot element for the νth step is $t_{r,s}^{(\nu-1)}$, then by Eq. (5.1-26) it is clear that

$$i=r: \quad b_r^{(\nu)} = \frac{b_r^{(\nu-1)}}{t_{r,s}^{(\nu-1)}};$$

$$i=1, 2, \ldots, r-1,$$

$$i=r+1, r+2, \ldots, m: \quad b_i^{(\nu)} = b_i^{(\nu-1)} - \frac{t_{i,s}^{(\nu-1)}}{t_{r,s}^{(\nu-1)}} b_r^{(\nu-1)}. \qquad (5.1\text{-}31)$$

The solution vector $\mathbf{y}^{(\nu)}$ at each step of the simplex iteration is a basic feasible solution that satisfies Eq. (5.1-27). Since it is a basic feasible solution vector, it has m nonzero components (unless it is a degenerate basic feasible solution, in which case it has less than m nonzero components). These nonzero components are the elements of $\mathbf{y}^{(\nu)}$ that, in the product $\mathbf{P}\mathbf{y}^{(\nu)}$, are the multipliers of those columns of $\mathbf{P}$ that constitute the basis corresponding to that basic feasible solution.

Now if $\mathbf{M}^{(\nu)}$ is a matrix whose columns are the basis vectors, in any order, then the tableau

$$\mathbf{T}^{(\nu)} = (\mathbf{M}^{(\nu)})^{-1}\,\mathbf{P}$$

is a matrix with m of its columns equal to the columns of the identity matrix, and those m columns occupy in $\mathbf{T}^{(\nu)}$ the same column locations as are occupied by the basis vectors in $\mathbf{P}$. Thus, the identity matrix columns are the ones that are multiplied by the nonzero components of $\mathbf{y}^{(\nu)}$ in the product $\mathbf{T}^{(\nu)}\mathbf{y}^{(\nu)}$, and the other columns of $\mathbf{T}^{(\nu)}$ are multiplied by zeros. As an example, consider again the solution and tableau that are obtained when the first iteration uses the matrix of basis vectors

$$\mathbf{M}^{(1)} = (\mathbf{p}_1, \mathbf{p}_{m+1}, \mathbf{p}_3, \ldots, \mathbf{p}_m).$$

The solution vector $\mathbf{y}^{(1)}$ will have its m nonzero elements in rows 1, 3, 4, . . ., m, m + 1, and Eq. (5.1-27) can then be written

$$\begin{bmatrix} 1 & t_{12}^{(1)} & 0 & \cdots & 0 & 0 & t_{1,m+2}^{(1)} & \cdots & t_{1,m+n}^{(1)} \\ 0 & t_{22}^{(1)} & 0 & \cdots & 0 & 1 & t_{2,m+2}^{(1)} & \cdots & t_{2,m+n}^{(1)} \\ 0 & t_{32}^{(1)} & 1 & \cdots & 0 & 0 & t_{3,m+2}^{(1)} & \cdots & t_{3,m+n}^{(1)} \\ \vdots & \vdots & \vdots & & \vdots & \vdots & \vdots & & \vdots \\ 0 & t_{m2}^{(1)} & 0 & \cdots & 1 & 0 & t_{m,m+2}^{(1)} & \cdots & t_{m,m+n}^{(1)} \end{bmatrix} \begin{bmatrix} y_1^{(1)} \\ 0 \\ y_3^{(1)} \\ \vdots \\ y_m^{(1)} \\ y_{m+1}^{(1)} \\ 0 \\ \vdots \\ 0 \end{bmatrix} = \begin{bmatrix} b_1^{(1)} \\ b_2^{(1)} \\ b_3^{(1)} \\ \vdots \\ b_m^{(1)} \end{bmatrix}$$

From this expression, it is clear that

$$y_1^{(1)} = b_1^{(1)},$$

$$y_2^{(1)} = 0,$$

$$y_j^{(1)} = b_j^{(1)}, \qquad j = 3, 4, \ldots, m.$$

$$y_{m+1}^{(1)} = b_2^{(1)},$$

$$y_j^{(1)} = 0, \qquad j = m+2, m+3, \ldots, m+n.$$

More generally, if the tableau $\mathbf{T}^{(\nu)}$ has columns $j_1, j_2, \ldots, j_m$ equal to the columns of the identity matrix; that is, if the basis matrix for the νth iteration is

$$\mathbf{M}^{(\nu)} = (\mathbf{p}_{j_1}, \mathbf{p}_{j_2}, \ldots, \mathbf{p}_{j_m}), \tag{5.1-32}$$

so that column j_1 of $\mathbf{T}^{(\nu)}$ is equal to the first column of the identity matrix $\mathbf{I}_m$, column j_2 of $\mathbf{T}^{(\nu)}$ is equal to the second column $\mathbf{I}_m$, and so on, then the elements of the solution vector $\mathbf{y}^{(\nu)}$ are given by

$$\begin{aligned} y_{j_i}^{(\nu)} &= b_i^{(\nu)}, \qquad i = 1, 2, \ldots, m, \\ y_j^{(\nu)} &= 0, \qquad j \neq j_i. \end{aligned} \tag{5.1-33}$$

Thus at each step of the iteration, the elements of the vector $\mathbf{b}^{(\nu)}$ and the location and arrangement of the identity matrix columns in $\mathbf{T}^{(\nu)}$ give all the information needed to write the solution $\mathbf{y}^{(\nu)}$.

The iterative procedure that has been described thus far provides a method for generating basic feasible solutions if the pivot at each step is chosen in such a way that none of the elements of the vector $\mathbf{b}^{(\nu)}$ become negative. Two things more are required to assure the success of the method. One is a criterion for determining whether or not each new solution generated is the maximum feasible solution.

The second is a strategy for choosing the pivots in such a way that each new iteration produces a larger value of the objective function, so that convergence to the maximum feasible solution is guaranteed without an excessive number of iterations. Therefore, it is necessary to take into account the vector $\mathbf{g}$, which defines the objective function

$$\phi(\mathbf{y}) = \mathbf{g}^T \mathbf{y}.$$

This vector is sometimes called the *cost vector,* a terminology that stems from the fact that most of the early applications of linear programming were concerned with economic or commercial problems (this terminology is perhaps more sensible when the linear programming problem is formulated as a minimization problem rather than as a maximization problem, as we have formulated it here).

It is convenient to consider the $(m+1) \times (n+m+1)$ composite matrix

$$\mathbf{Q} = \begin{bmatrix} \mathbf{P} & \mathbf{b} \\ -\mathbf{g}^T & 0 \end{bmatrix} \tag{5.1-34}$$

and sequences of matrices of the form

$$\mathbf{Q}^{(\nu)} = \begin{bmatrix} \mathbf{T}^{(\nu)} & \mathbf{b}^{(\nu)} \\ -(\mathbf{g}^{(\nu)})^T & \gamma_\nu \end{bmatrix} \tag{5.1-35}$$

where each $\mathbf{Q}^{(\nu)}$ is obtainable from $\mathbf{Q}$ by performing a series of elementary row operations on $\mathbf{Q}$, with the restriction that the only operations that are allowed on the last row are additions of scalar multiples of the other rows. It can be shown [2, pp. 163-164] that the linear programming problem defined by

$$\phi^{(\nu)} = \max_{\mathbf{y} \geq \mathbf{o}} \{\gamma_\nu + (\mathbf{g}^{(\nu)})^T \mathbf{y} | \mathbf{T}^{(\nu)} \mathbf{y} = \mathbf{b}^{(\nu)}\} \tag{5.1-36}$$

is completely equivalent to the original problem

$$\phi_0 = \max_{\mathbf{y} \geq \mathbf{o}} \{\mathbf{g}^T \mathbf{y} | \mathbf{P}\mathbf{y} = \mathbf{b}\}.$$

We will refer to the matrices $\mathbf{Q}^{(\nu)}$ as the *extended tableaux.* The matrices $\mathbf{T}^{(\nu)}$ and the vectors $\mathbf{b}^{(\nu)}$ will be the by now familiar ones defined by Eqs. (5.1-20) and (5.1-28). The vector $\mathbf{g}^{(\nu)}$ will be constructed in such a way that its elements provide the necessary information to establish whether or not the iteration has converged and so that the new objective function

$$\phi^{(\nu)}(\mathbf{y}) = \gamma_\nu + (\mathbf{g}^{(\nu)})^T \mathbf{y} \tag{5.1-37}$$

has the value γ_ν when $\mathbf{y}$ is equal to the basic feasible solution vector $\mathbf{y}^{(\nu)}$; that is, so that

$$(\mathbf{g}^{(\nu)})^T \mathbf{y}^{(\nu)} = 0. \tag{5.1-38}$$

We have already seen how the matrix $\mathbf{T}^{(0)}$ is obtained by the Gauss-Jordan reduction and how the vector $\mathbf{b}^{(0)}$ is obtained by extending the same elementary row operations to $\mathbf{b}$. The vector $(\mathbf{g}^{(0)})^T$ has the form

$$(\mathbf{g}^{(0)})^T = (0, 0, \ldots, 0, g_{m+1}^{(0)}, g_{m+2}^{(0)}, \ldots, g_{m+n}^{(0)})$$

and is obtained by extending the Gauss-Jordan reduction to the bottom row of the matrix $\mathbf{Q}$, at each step choosing the row operation to annihilate the element of $\mathbf{g}$ that is in the same column as the pivot element for that step. Let

$$[-(\mathbf{g}^{[0]})^T, \gamma^{[0]}] = (-\mathbf{g}^T, 0),$$

$$[-(\mathbf{g}^{[1]})^T, \gamma^{[1]}],$$

$$[-(\mathbf{g}^{[2]})^T, \gamma^{[2]}],$$

$$\ldots\ldots\ldots\ldots\ldots\ldots\ldots,$$

$$[-(\mathbf{g}^{[m]})^T, \gamma^{[m]}] = [-(\mathbf{g}^{(0)})^T, \gamma_0],$$

denote the bottom row of the transformed **Q** matrix at each step of the initial Gauss-Jordan reduction. The equations for transforming the bottom row in the first step of the reduction are

$$-g_j^{[1]} = -g_j^{[0]} - \frac{-g_1^{[0]}}{p_{11}^{(0)}} p_{1j}^{(0)}, \qquad j = 1, 2, \ldots, m+n,$$

$$\gamma^{[1]} = 0 - \frac{-g_1^{[0]}}{p_{11}^{(0)}} b_1^{[0]},$$

and the matrix **Q** is transformed into

$$\left[\begin{array}{ccccc|c} 1 & p_{12}^{(1)} & p_{13}^{(1)} & \cdots & p_{1,m+n}^{(1)} & b_1^{[1]} \\ 0 & p_{22}^{(1)} & p_{23}^{(1)} & \cdots & p_{2,m+n}^{(1)} & b_2^{[1]} \\ \vdots & \vdots & \vdots & & \vdots & \vdots \\ 0 & p_{m2}^{(1)} & p_{m3}^{(1)} & \cdots & p_{m,m+n}^{(1)} & b_m^{[1]} \\ \hline 0 & -g_2^{[1]} & -g_3^{[1]} & \cdots & -g_{m+n}^{[1]} & \gamma^{[1]} \end{array}\right]$$

where the $p_{ij}^{(1)}$ are defined by the transformations (5.1-22) and the $b_i^{[1]}$ by (5.1-30) with $k = 1$. Each subsequent step of the reduction produces a new column of the identity matrix in the first m rows with a zero directly below it in the bottom row. The equations for the transformations on the bottom row at the kth step of the reduction are

$$-g_j^{[k]} = -g_j^{[k-1]} - \frac{-g_k^{[k-1]}}{p_{kk}^{[k-1]}} p_{kj}^{[k-1]}, \qquad j = 1, 2, \ldots, m+n,$$

$$\gamma^{[k]} = \gamma^{[k-1]} - \frac{-g_k^{[k-1]}}{p_{kk}^{[k-1]}} b_k^{[k-1]}. \tag{5.1-39}$$

The result of m such steps is the initial extended tableau

$$Q^{(0)} = \left[\begin{array}{cccccccc|c} 1 & 0 & \cdots & 0 & t^{(0)}_{1,m+1} & \cdots & t^{(0)}_{1,m+n} & & b^{(0)}_1 \\ 0 & 1 & \cdots & 0 & t^{(0)}_{2,m+1} & \cdots & t^{(0)}_{2,m+n} & & b^{(0)}_2 \\ \vdots & \vdots & & \vdots & \vdots & & \vdots & & \vdots \\ 0 & 0 & \cdots & 1 & t^{(0)}_{m,m+1} & \cdots & t^{(0)}_{m,m+n} & & b^{(0)}_m \\ \hline 0 & 0 & \cdots & 0 & -g^{(0)}_{m+1} & \cdots & -g^{(0)}_{m+n} & & \gamma_0 \end{array}\right]. \quad (5.1\text{-}40)$$

The vector

$$-\mathbf{g}^{(0)} = (0, 0, \ldots, 0, -g^{(0)}_{m+1}, \ldots, -g^{(0)}_{m+n})^T$$

could also be computed from the formula

$$\mathbf{g}^{(0)} = \mathbf{g} - [(\mathbf{M}^{(0)})^{-1}\mathbf{P}]^T\mathbf{h}^{(0)} = \mathbf{g} - (\mathbf{T}^{(0)})^T\mathbf{h}^{(0)} \quad (5.1\text{-}41)$$

where the m-vector $\mathbf{h}^{(0)}$ is defined by

$$\mathbf{h}^{(0)} = (g_1, g_2, \ldots, g_m)^T. \quad (5.1\text{-}42)$$

Similarly, if

$$\mathbf{M}^{(\nu)} = (\mathbf{p}_{j_1}, \mathbf{p}_{j_2}, \ldots, \mathbf{p}_{j_m}) \quad (5.1\text{-}43)$$

is the basis matrix for the νth iteration, then the vector $\mathbf{g}^{(\nu)}$ could be computed from

$$\mathbf{g}^{(\nu)} = \mathbf{g} - [(\mathbf{M}^{(\nu)})^{-1}\mathbf{P}]^T\mathbf{h}^{(\nu)} = \mathbf{g} - (\mathbf{T}^{(\nu)})^T\mathbf{h}^{(\nu)} \quad (5.1\text{-}44)$$

where

$$\mathbf{h}^{(\nu)} = (g_{j_1}, g_{j_2}, \ldots, g_{j_m})^T. \quad (5.1\text{-}45)$$

If the pivot element for the νth iteration is $t^{(\nu-1)}_{r,s}$, then in actual practice, $\mathbf{g}^{(\nu)}$ is computed by

$$-g^{(\nu)}_j = -g^{(\nu-1)}_j - \frac{-g^{(\nu-1)}_s}{t^{(\nu-1)}_{r,s}}\, t^{(\nu-1)}_{r,j}, \qquad j = 1. 2, \ldots, m+n. \quad (5.1\text{-}46)$$

One result of this transformation is to annihilate the element $g_s^{(\nu-1)}$ and at the same time to produce a nonzero value $g_r^{(\nu)}$ in place of an element that was formerly equal to zero. For example, if the first iteration uses the element $t_{2,m+1}^{(0)}$ as the pivot element, then $\mathbf{Q}^{(1)}$ has the form

$$\mathbf{Q}^{(1)} = \left[\begin{array}{cccccccccc|c} 1 & t_{12}^{(1)} & 0 & \cdots & 0 & 0 & t_{1,m+2}^{(1)} & \cdots & t_{1,m+n}^{(1)} & & b_1^{(1)} \\ 0 & t_{22}^{(1)} & 0 & \cdots & 0 & 1 & t_{2,m+2}^{(1)} & \cdots & t_{2,m+n}^{(1)} & & b_2^{(1)} \\ 0 & t_{32}^{(1)} & 1 & \cdots & 0 & 0 & t_{3,m+2}^{(1)} & \cdots & t_{3,m+n}^{(1)} & & b_3^{(1)} \\ \vdots & \vdots & \vdots & & \vdots & \vdots & \vdots & & \vdots & & \vdots \\ 0 & t_{m2}^{(1)} & 0 & \cdots & 1 & 0 & t_{m,m+2}^{(1)} & \cdots & t_{m,m+n}^{(1)} & & b_m^{(1)} \\ \hline 0 & g_2^{(1)} & 0 & \cdots & 0 & 0 & g_{m+2}^{(1)} & \cdots & g_{m+n}^{(1)} & & \gamma_1 \end{array}\right].$$

In this case, the element γ_1 is obtained by

$$\gamma_1 = \gamma_0 - \frac{-g_{m+1}^{(0)}}{t_{2,m+1}^{(0)}} b_2^{(0)}.$$

In the general case, when the pivot is $t_{r,s}^{(\nu-1)}$, γ_ν is obtained by

$$\gamma_\nu = \gamma_{\nu-1} - \frac{-g_s^{(\nu-1)}}{t_{r,s}^{(\nu-1)}} b_r^{(\nu-1)}. \tag{5.1-47}$$

The vector $\mathbf{g}^{(\nu)}$ is sometimes called the *relative cost vector.* At each step of the iteration, the new objective function is

$$\phi^{(\nu)}(\mathbf{y}) = \gamma_\nu + (\mathbf{g}^{(\nu)})^T \mathbf{y},$$

so that

$$\frac{\partial \phi^{(\nu)}(\mathbf{y})}{\partial \mathbf{y}} = \mathbf{g}^{(\nu)}. \tag{5.1-48}$$

Therefore, the vector $\mathbf{g}^{(\nu)}$ tells at each step how the objective function varies with changes in the vector $\mathbf{y}$. The only nonzero elements in $\mathbf{g}^{(\nu)}$ are the components that correspond to the columns of $\mathbf{P}$ that are not in the basis; that is, that correspond to the columns of $\mathbf{T}^{(\nu)}$ that are not equal to identity matrix columns.

Hence, the nonzero elements of $\mathbf{g}^{(\nu)}$ are the ones that correspond to the nonbasic variables for the νth step of the iteration. This means that

$$(\mathbf{g}^{(\nu)})^T \mathbf{y}^{(\nu)} = 0,$$

so that

$$\phi^{(\nu)}(\mathbf{y}^{(\nu)}) = \gamma_\nu. \tag{5.1-49}$$

Thus, at any step ν of the iteration, γ_ν is equal to the value of the objective function given by the basic feasible solution $\mathbf{y}^{(\nu)}$ at that step. If the basic variables for the νth iteration are $y_{j_1}, y_{j_2}, \ldots, y_{j_m}$ and the nonbasic variables are $y_{j_{m+1}}, y_{j_{m+2}}, \ldots, y_{j_{m+n}}$, then the objective function (5.1-37) can be written

$$\phi^{(\nu)}(\mathbf{y}) = \gamma_\nu + g_{j_{m+1}}^{(\nu)} y_{j_{m+1}} + g_{j_{m+2}}^{(\nu)} y_{j_{m+2}} + \ldots + g_{j_{m+n}}^{(\nu)} y_{j_{m+n}}. \tag{5.1-50}$$

Clearly then, changes in the objective function can be effected only by changes in the nonbasic variables. For the solution vector $\mathbf{y}^{(\nu)}$, the nonbasic variables are all equal to zero and the only changes allowed by the nonnegativity constraint are to make one or more of them positive. In the simplex algorithm, the effect of the elimination calculation at each step is to increase one of the nonbasic variables, leaving the others equal to zero, and making the changes in the basic variables that are required to maintain the constraint

$$\mathbf{P}\mathbf{y} = \mathbf{b}.$$

In the process, one of the basic variables is annihilated and thus becomes a nonbasic variable. If all of the nonzero elements of $\mathbf{g}^{(\nu)}$ are negative, then it is clear from (5.1-50) that an increase in any of the nonbasic variables will produce a decrease in the objective function. Thus, at any stage ν of the iteration, if all of the components of $\mathbf{g}^{(\nu)}$ that correspond to the nonbasic variables for that stage are strictly negative; that is, if

$$g_{j_{m+1}}^{(\nu)}, g_{j_{m+2}}^{(\nu)}, \ldots, g_{j_{m+n}}^{(\nu)} < 0,$$

then the iteration has converged and the vector $\mathbf{y}^{(\nu)}$ is the unique maximum feasible solution. The maximum value of the objective function is given by

$$\phi_0 = \phi^{(\nu)}(\mathbf{y}^{(\nu)}) = \gamma_\nu.$$

If, at any step ν, all of the components of $\mathbf{g}^{(\nu)}$ corresponding to nonbasic variables become nonpositive, that is,

$$g_{j_{m+1}}^{(\nu)}, g_{j_{m+2}}^{(\nu)}, \ldots, g_{j_{m+n}}^{(\nu)} \leqslant 0,$$

but with one or more of them being exactly equal to zero, then the current solution vector $\mathbf{y}^{(\nu)}$ is a maximum feasible solution, but it is not unique. The maximum value of the objective function is still

$$\phi_0 = \gamma_\nu,$$

but there exists at least one other basic feasible solution vector for which $\phi(\mathbf{y})$ achieves its maximum value. Furthermore, if $\phi(\mathbf{y})$ is maximized by more than one basic feasible solution vector, then any convex combination of these vectors will also produce the maximum value $\phi_0 = \gamma_\nu$. This is the situation that is illustrated graphically in Fig. 5.4.

If at least one of the nonzero components of $\mathbf{g}^{(\nu)}$ is positive, then it is possible to increase the value of $\phi^{(\nu)}(\mathbf{y})$ by increasing the corresponding nonbasic variable. For the sake of definiteness, consider the initial extended tableau

$$\mathbf{Q}^{(0)} = \left[\begin{array}{cccccccc|c} 1 & 0 & \cdots & 0 & t_{1,m+1}^{(0)} & \cdots & t_{1,m+n}^{(0)} & & b_1^{(0)} \\ 0 & 1 & \cdots & 0 & t_{2,m+1}^{(0)} & \cdots & t_{2,m+n}^{(0)} & & b_2^{(0)} \\ \vdots & \vdots & & \vdots & \vdots & & \vdots & & \vdots \\ 0 & 0 & \cdots & 1 & t_{m,m+1}^{(0)} & \cdots & t_{m,m+n}^{(0)} & & b_m^{(0)} \\ \hline 0 & 0 & \cdots & 0 & -g_{m+1}^{(0)} & \cdots & -g_{m+n}^{(0)} & & \gamma_0 \end{array}\right]$$

and the corresponding basic feasible solution vector

$$\begin{aligned} \mathbf{y}^{(0)} &= (y_1^{(0)}, y_2^{(0)}, \ldots, y_m^{(0)}, 0, \ldots, 0)^T \\ &= (b_1^{(0)}, b_2^{(0)}, \ldots, b_m^{(0)}, 0, \ldots, 0)^T. \end{aligned}$$

The new objective function

$$\phi^{(0)}(\mathbf{y}) = \gamma_0 + g_{m+1}^{(0)} y_{m+1} + \ldots + g_{m+n}^{(0)} y_{m+n}$$

has the value γ_0 when $\mathbf{y} = \mathbf{y}^{(0)}$. Now suppose that $g_{m+1}^{(0)}$ is positive. Then the value of $\phi^{(0)}(\mathbf{y})$ can be increased by increasing y_{m+1}. Therefore, for the next step of the simplex iteration, we might choose to make y_{m+1} one of the basic variables, at the same time making one of $y_1, y_2, \ldots, y_m$ a nonbasic variable. Another way to say this is that we might choose the pivot for the next step to be in the $(m+1)$st column of $\mathbf{T}^{(0)}$.

To determine the row for the pivot element, we must consider the effect of the elements $t_{1,m+1}^{(0)}, t_{2,m+1}^{(0)}, \ldots, t_{m,m+1}^{(0)}$ and of the elements of the vector $\mathbf{b}^{(0)}$ upon $y_1, y_2, \ldots, y_m$ whenever changes are made in y_{m+1}. Any increases in y_{m+1} must be accompanied by changes in $y_1, y_2, \ldots, y_m$ in order to maintain the constraint

$$\mathbf{Py} = \mathbf{b}$$

or, equivalently

$$\mathbf{T}^{(0)}\mathbf{y} = \mathbf{b}^{(0)}.$$

If none of the other variables $y_{m+2}, \ldots, y_{m+n}$ are allowed to change, then this constraint can be written

$$\begin{bmatrix} 1 & 0 & \cdots & 0 & t_{1,m+1}^{(0)} & t_{1,m+2}^{(0)} & \cdots & t_{1,m+n}^{(0)} \\ 0 & 1 & \cdots & 0 & t_{2,m+1}^{(0)} & t_{2,m+2}^{(0)} & \cdots & t_{2,m+n}^{(0)} \\ \vdots & \vdots & & \vdots & \vdots & \vdots & & \vdots \\ 0 & 0 & \cdots & 1 & t_{m,m+1}^{(0)} & t_{m,m+2}^{(0)} & \cdots & t_{m,m+n}^{(0)} \end{bmatrix} \begin{bmatrix} y_1 \\ y_2 \\ \vdots \\ y_m \\ y_{m+1} \\ 0 \\ \vdots \\ 0 \end{bmatrix} = \begin{bmatrix} b_1^{(0)} \\ b_2^{(0)} \\ \vdots \\ b_m^{(0)} \end{bmatrix},$$

from which it is readily apparent that

$$y_i + t_{i,m+1}^{(0)} y_{m+1} = b_i^{(0)}, \qquad i = 1, 2, \ldots, m,$$

or

$$y_i = b_i^{(0)} - t_{i,m+1}^{(0)} y_{m+1}, \qquad i = 1, 2, \ldots, m. \qquad (5.1\text{-}51)$$

At any step ν of the simplex iteration, the $b_i^{(\nu)}$ are the nonzero components of the basic feasible solution vector for that step, so they must all be nonnegative. In fact, they must be strictly positive if the solution is a nondegenerate basic feasible solution. We assume that $\mathbf{y}^{(0)}$ is such a nondegenerate solution, so that all the $b_i^{(0)} = y_i^{(0)}$ are strictly positive. Now y_{m+1} may be increased only so long as $y_1, y_2, \ldots, y_m$ all remain nonnegative. Thus, if none of the $t_{i,m+1}^{(0)}$ is positive, then it is possible to find feasible solutions that make the objective function $\phi^{(0)}(\mathbf{y})$ as large as we please. This is the case, illustrated in Fig. 5.3, where the constraint region is unbounded in some direction and the objective function increases in that direction.

If at least one of the $t_{i,m+1}^{(0)}$ is positive, there is a limit on how much y_{m+1} may be increased. It is clear from Eq. (5.1-51) that if $t_{k,m+1}^{(0)} > 0$, then y_k becomes negative when y_{m+1} exceeds the value $b_k^{(0)}/t_{k,m+1}^{(0)}$. If several of the $t_{i,m+1}^{(0)}$ are positive, then y_{m+1} cannot exceed the minimum value defined by

$$\frac{b_r^{(0)}}{t_{r,m+1}^{(0)}} = \min_i \left\{ \frac{b_i^{(0)}}{t_{i,m+1}^{(0)}} \middle| t_{i,m+1}^{(0)} > 0 \right\}. \tag{5.1-52}$$

Since we want to increase the value of the objective function as much as possible, we would like to take

$$y_{m+1}^{(1)} = \frac{b_r^{(0)}}{t_{r,m+1}^{(0)}}.$$

This is exactly what happens in the next iteration of the simplex method if we choose $t_{r,m+1}^{(0)}$ to be the pivot element. From (5.1-51), it is clear that in the process, the variable y_r is annihilated; that is, $y_r^{(1)} = 0$, and thus becomes a nonbasic variable. It is also clear that the new basic variables $y_1^{(1)}, \ldots, y_{r-1}^{(1)}, y_{r+1}^{(1)}, \ldots, y_m^{(1)}, y_{m+1}^{(1)}$ are all positive. In the actual calculations, we work with the elements of the vectors $\mathbf{b}^{(0)}$ and $\mathbf{b}^{(1)}$, and the strategy just described for choosing the pivot row guarantees that the elements of $\mathbf{b}^{(1)}$ will all be nonnegative.

Now suppose that after the νth iteration of the simplex algorithm, the basic variables are $y_{j_1}, y_{j_2}, \ldots, y_{j_m}$, and the nonbasic variables are $y_{j_{m+1}}, y_{j_{m+2}}, \ldots, y_{j_{m+n}}$. If the components of the relative cost vector $\mathbf{g}^{(\nu)}$ satisfy

$$g_{j_{m+1}}^{(\nu)}, g_{j_{m+2}}^{(\nu)}, \ldots, g_{j_{m+n}}^{(\nu)} \leqslant 0,$$

then $\mathbf{y}^{(\nu)}$ is a maximum feasible solution (it is the unique maximum feasible solution if all the foregoing components are strictly negative), and the problem is solved. If one or more of $g_{j_{m+1}}^{(\nu)}, g_{j_{m+2}}^{(\nu)}, \ldots, g_{j_{m+n}}^{(\nu)}$ is positive, then $\mathbf{y}^{(\nu)}$ is not a maximum feasible solution, and the objective function can be increased by iterating again with a pivot chosen in one of the columns of $\mathbf{T}^{(\nu)}$ corresponding to a positive $g_j^{(\nu)}$. If for any such column all the entries are nonpositive, that is, if

$$g_j^{(\nu)} > 0 \quad \text{and} \quad t_{ij}^{(\nu)} \leqslant 0, \qquad i = 1, 2, \ldots, m,$$

then the objective function is unbounded in the given constraint region. In this case, the problem does not have a finite maximum feasible solution and there is no point in continuing the iteration. But if all of the columns corresponding to a positive $g_j^{(\nu)}$ have at least one positive entry, then the iteration should be continued. The usual strategy for choosing the pivot column is to pick the column s for which

$$g_s^{(\nu)} = \max_j \{g_j^{(\nu)} \mid g_j^{(\nu)} > 0\}. \tag{5.1-53}$$

This is not always the best possible choice, but it has the advantage that it produces the maximum change in the objective function per unit change in the corresponding variable y_s. Once the pivot column has been determined, the pivot row is chosen to be that row r for which

$$\frac{b_r^{(\nu)}}{t_{r,s}^{(\nu)}} = \min_i \left\{ \frac{b_i^{(\nu)}}{t_{i,s}^{(\nu)}} \middle| t_{i,s}^{(\nu)} > 0 \right\}. \tag{5.1-54}$$

This choice of $t_{r,s}^{(\nu)}$ as the pivot replaces y_r by y_s as a basic variable and produces the maximum increase in the objective function that is consistent with the nonnegativity constraint and with the choice of column s as the pivot column.

Figure 5.5 gives a schematic outline of the simplex method as it has been developed in this section. Box 1 contains the original statement of the problem and box 2 contains the restatement when slack variables are added. Box 3 contains the initial extended tableau $\mathbf{Q}^{(0)}$, which can be obtained by either of two methods. In actual practice, it is, of course, much easier to simply rewrite the problem with the slack variables first; that is, to take

$$\mathbf{y} = \begin{pmatrix} \mathbf{s} \\ \mathbf{x} \end{pmatrix}, \quad \mathbf{g} = \begin{pmatrix} \mathbf{0} \\ \mathbf{c} \end{pmatrix}, \quad \text{and} \quad \mathbf{T}^{(0)} = (\mathbf{I}_m, \mathbf{A}),$$

than to carry out the extensive calculations involved in the Gauss-Jordan reduction. In fact, there is actually no need even to rewrite the problem, since the initial tableau can be taken to be

$$\mathbf{T}^{(0)} = (\mathbf{A}, \mathbf{I}_m)$$

just as well as

$$\mathbf{T}^{(0)} = (\mathbf{I}_m, \mathbf{A}).$$

The only real requirement is that $\mathbf{T}^{(0)}$ have m of its columns equal to the columns of the identity matrix in some order. Although this method is the easier way to start the iteration, it was instructive to introduce the Gauss-Jordan reduction because in each step of the simplex procedure, the tableau $\mathbf{T}^{(\nu)}$ is obtained from $\mathbf{T}^{(\nu-1)}$ by an elimination procedure that is computationally identical to one step of the Gauss-Jordan reduction. Furthermore, some linear programming problems are formulated with equality constraints of the form

$$\mathbf{Ax} = \mathbf{b},$$

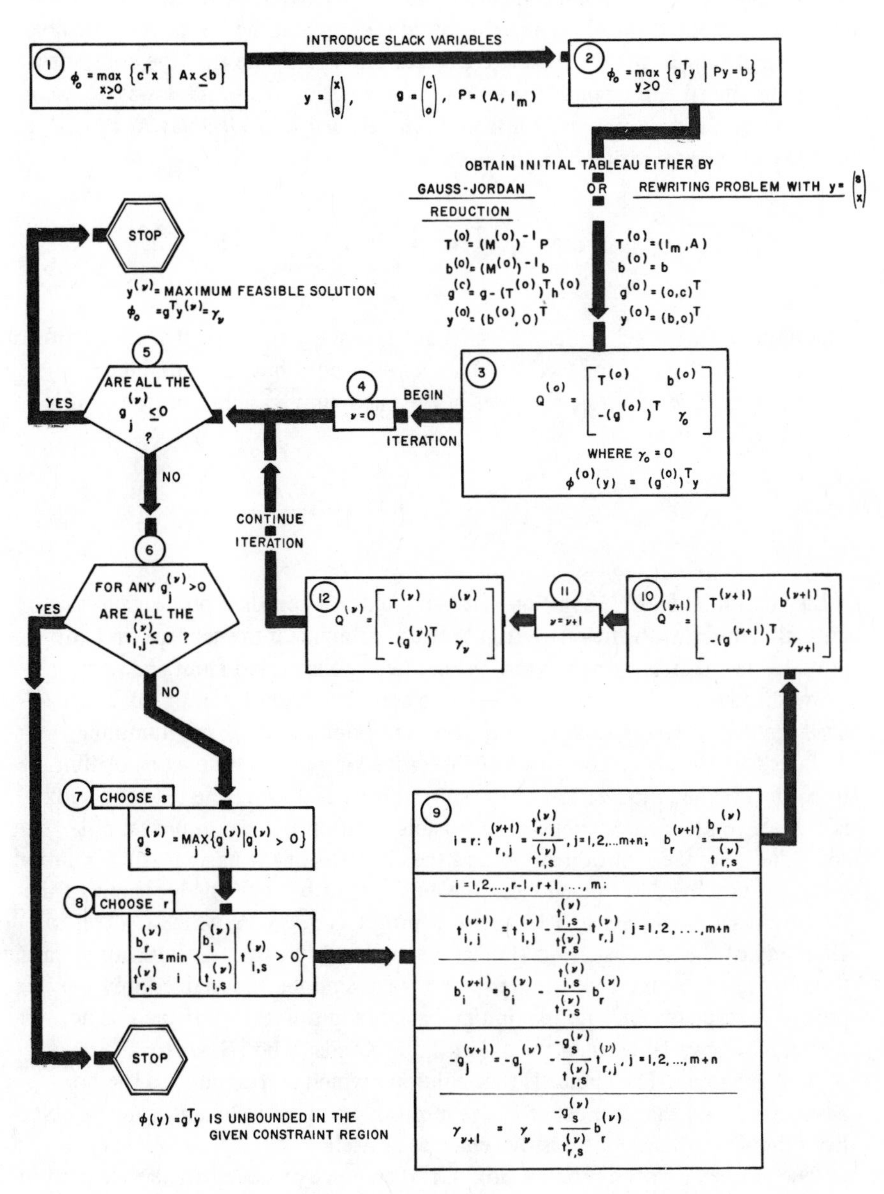

INTRODUCE SLACK VARIABLES
1 $\phi_o = \max_{x \geq 0} \{c^T x \mid Ax \leq b\}$
2 $\phi_o = \max_{y \geq 0} \{g^T y \mid Py = b\}$
$y = \begin{pmatrix} x \\ s \end{pmatrix}$, $g = \begin{pmatrix} c \\ o \end{pmatrix}$, $P = (A, I_m)$
OBTAIN INITIAL TABLEAU EITHER BY
GAUSS-JORDAN REDUCTION
OR
REWRITING PROBLEM WITH $y = \begin{pmatrix} s \\ x \end{pmatrix}$
$T^{(o)} = (M^{(o)})^{-1} P$
$b^{(o)} = (M^{(o)})^{-1} b$
$g^{(c)} = g - (T^{(o)})^T h^{(o)}$
$y^{(o)} = (b^{(o)}, 0)^T$
$T^{(o)} = (I_m, A)$
$b^{(o)} = b$
$g^{(o)} = (o, c)^T$
$y^{(o)} = (b, o)^T$
STOP
$y^{(\nu)}$ = MAXIMUM FEASIBLE SOLUTION
$\phi_o = g^T y^{(\nu)} = \gamma_\nu$
5 ARE ALL THE $g_j^{(\nu)} \leq 0$?
YES
NO
4 $\nu = 0$
BEGIN ITERATION
3 $Q^{(o)} = \begin{bmatrix} T^{(o)} & b^{(o)} \\ -(g^{(o)})^T & \gamma_o \end{bmatrix}$
WHERE $\gamma_o = 0$
$\phi^{(o)}(y) = (g^{(o)})^T y$
CONTINUE ITERATION
6 FOR ANY $g_j^{(\nu)} > 0$ ARE ALL THE $t_{i,j}^{(\nu)} \leq 0$?
YES
NO
12 $Q^{(\nu)} = \begin{bmatrix} T^{(\nu)} & b^{(\nu)} \\ -(g^{(\nu)})^T & \gamma_\nu \end{bmatrix}$
11 $\nu = \nu + 1$
10 $Q^{(\nu+1)} = \begin{bmatrix} T^{(\nu+1)} & b^{(\nu+1)} \\ -(g^{(\nu+1)})^T & \gamma_{\nu+1} \end{bmatrix}$
7 CHOOSE s
$g_s^{(\nu)} = \mathrm{MAX}_j \{g_j^{(\nu)} \mid g_j^{(\nu)} > 0\}$
8 CHOOSE r
$\frac{b_r^{(\nu)}}{t_{r,s}^{(\nu)}} = \min_i \left\{ \frac{b_i^{(\nu)}}{t_{i,s}^{(\nu)}} \middle| t_{i,s}^{(\nu)} > 0 \right\}$
STOP
$\phi(y) = g^T y$ IS UNBOUNDED IN THE GIVEN CONSTRAINT REGION
9 $i = r$: $t_{r,j}^{(\nu+1)} = \frac{t_{r,j}^{(\nu)}}{t_{r,s}^{(\nu)}}$, $j = 1, 2, \ldots m+n$; $b_r^{(\nu+1)} = \frac{b_r^{(\nu)}}{t_{r,s}^{(\nu)}}$
$i = 1, 2, \ldots, r-1, r+1, \ldots, m$:
$t_{i,j}^{(\nu+1)} = t_{i,j}^{(\nu)} - \frac{t_{i,s}^{(\nu)}}{t_{r,s}^{(\nu)}} t_{r,j}^{(\nu)}$, $j = 1, 2, \ldots, m+n$
$b_i^{(\nu+1)} = b_i^{(\nu)} - \frac{t_{i,s}^{(\nu)}}{t_{r,s}^{(\nu)}} b_r^{(\nu)}$
$-g_j^{(\nu+1)} = -g_j^{(\nu)} - \frac{-g_s^{(\nu)}}{t_{r,s}^{(\nu)}} t_{r,j}^{(\nu)}$, $j = 1, 2, \ldots, m+n$
$\gamma_{\nu+1} = \gamma_\nu - \frac{-g_s^{(\nu)}}{t_{r,s}^{(\nu)}} b_r^{(\nu)}$

Figure 5.5

so there are no slack variables. In such cases, the Gauss–Jordan reduction is a convenient method for obtaining the initial tableau and the initial basic feasible solution, although care must be exercised in order to assure that the columns that are reduced correspond to a nonnegative solution vector. One way to accomplish this aim is to temporarily introduce m new *artificial variables* $z_1, z_2, \ldots, z_m$ into the problem, writing it in the form

$$\phi_0 = \max_{(\mathbf{x},\mathbf{z}) \geqslant \mathbf{o}} \left\{ (\mathbf{c}^T, \boldsymbol{\epsilon}^T) \begin{pmatrix} \mathbf{x} \\ \mathbf{z} \end{pmatrix} \middle| (\mathbf{A}, \mathbf{I}_m) \begin{pmatrix} \mathbf{x} \\ \mathbf{z} \end{pmatrix} = \mathbf{b} \right\}$$

where the m-vector $\boldsymbol{\epsilon}$ has elements that are extremely small, so that they contribute essentially nothing to the objective function. Thus, the variables $z_1, z_2, \ldots, z_m$ will all be nonbasic variables in the maximum feasible solution to this augmented problem. The vector

$$\begin{pmatrix} \mathbf{x}^{(0)} \\ \mathbf{z}^{(0)} \end{pmatrix} = \begin{pmatrix} \mathbf{0} \\ \mathbf{b} \end{pmatrix}$$

is then an initial feasible solution. The simplex iteration then proceeds in the normal manner, with the restriction that the columns of the tableau that correspond to the artificial variables are systematically eliminated from and never allowed to return to the basis set. A more complete description of this *artificial-basis technique* can be found in any good textbook on linear programming.

Boxes 5 through 12 constitute the iterative loop of the simplex algorithm. Box 4 is just the initialization of the subscript ν before entering the loop. Box 5 is the test to see if the current basic feasible solution is a maximum feasible solution. If so, the iteration stops, and the elements of the basic feasible solution vector $\mathbf{y}^{(\nu)}$ can be determined from the elements of $\mathbf{b}^{(\nu)}$ and the locations of the identity matrix columns in $\mathbf{T}^{(\nu)}$; if not, it proceeds to box 6, which is a test to see if any of the nonbasic variables can be made arbitrarily large without violating the constraints on the problem. If there is such a nonbasic variable, then the process is stopped, since no maximum feasible solution exists. If there is no such nonbasic variable, then the algorithm proceeds to box 7, where the pivot column is chosen. The choice is that column s which corresponds to the nonbasic variable y_s that is most effective in producing positive changes in the objective function by means of positive changes in itself.

The next step of the process, box 8, is the choice of the particular element in column s to be used as the pivot. This choice is the element $t_{r,s}^{(\nu)}$ that permits the maximum increase in $\phi^{(\nu)}(\mathbf{y})$ and at the same time preserves the nonnegativity constraints on all the basic variables. At this point, we should note that if the current basic feasible solution is a degenerate solution; that is, if one or more of the $b_i^{(\nu)}$ are zero, then the pivot will be chosen in a row corresponding to a zero value; that is, r will be chosen so that $b_r^{(\nu)} = 0$. This means that

$$\phi(\mathbf{y}^{(\nu+1)}) = \gamma_{\nu+1} = \gamma_\nu - \frac{-g_s^{(\nu)}}{t_{r,s}^{(\nu)}} b_r^{(\nu)}$$

$$= \gamma_\nu = \phi(\mathbf{y}^{(\nu)}).$$

Thus the objective function will not be changed by the next step of the iteration. Such an occurence in isolated incidents presents no difficulties; but if the same thing should happen for several successive steps, the algorithm might repeat a previous basic feasible solution and then start regenerating the same series of solutions over and over again without any improvement in the objective function. This phenomenon is called *cycling.* It cannot occur in any problem for which the $m \times (m+n+1)$ composite matrix

$$(\mathbf{P}, \mathbf{b})$$

has the property that every $m \times m$ submatrix selected from it is nonsingular, the reason being that such a problem does not have any degenerate basic feasible solutions [2, pp. 164–165]. Most authors on the subject of linear programming report that although degenerate solutions are very common, cycling almost never occurs; but it has been the experience of the present authors that cycling is a fairly frequent occurrence for the linear programming problems that arise from applying the constrained estimation techniques discussed in Chapter 4 to singular and poorly conditioned systems of equations.

There are two widely known techniques for avoiding cycling in the simplex algorithm. One of them is a perturbation technique that was developed by Charnes. It is based upon the idea of making small alterations in the elements of the vector **b** so that when the perturbed **b** is written as a linear combination of any m basis vectors chosen from the columns of **P**, a positive amount of each of the basis vectors is required. Thus, the perturbations prevent the occurrence of any degenerate basic feasible solutions. The other technique is the generalized simplex method developed by Dantzig, Orden, and Wolfe. It is based upon the idea of lexicographic ordering of vectors. A good discussion of these two methods has been given by Hadley [3, Chapter 6].

Once the pivot element $t_{r,s}^{(\nu)}$ has been chosen, it is a simple matter to compute the new extended tableau $\mathbf{Q}^{(\nu+1)}$ by the equations given in box 9. Box 10 is just a symbolic representation of this new tableau. Box 11 is a Fortran-like statement that means that the next iteration is about to begin with all the superscripts increased by one, and box 12 is another symbolic representation of the new tableau with the superscripts renamed for the next iteration. If cycling does not occur, then the algorithm keeps iterating until, after a finite number of steps, a maximum feasible solution is obtained (or else until it becomes apparent that no such solution exists). The number of steps required is usually somewhere between m and $2m$.

The simplex method, as it has been outlined here, provides an algorithm for calculating the maximum feasible solution corresponding to a given matrix **A** and given vectors **b** and **c**. But it is often desirable to know how the maximum feasible solution varies as the elements of the vector **b** vary continuously from one value to another. This kind of "sensitivity analysis" can be carried out by a technique called *parametric linear programming.* An excellent discussion of this procedure has been given by Garvin [4, Chapter 15]. We will consider here only the case where the elements of the vector **b** are all linear functions of some nonnegative parameter θ; that is, the case

$$\mathbf{b}(\theta) = \mathbf{b}(0) + \theta\boldsymbol{\beta}, \qquad \theta \geqslant 0, \tag{5.1-55}$$

where $\mathbf{b}(0)$ is some given fixed vector corresponding to $\theta = 0$ and $\boldsymbol{\beta}$ is some fixed vector of constants (positive, negative, or zero) that tell how the variations in the elements $b_i(\theta)$ are related to one another. We again consider the problem after it has been augmented by the slack variables; that is, the problem

$$\phi_0 = \max_{\mathbf{y}(\theta)\geqslant\mathbf{o}} \{\mathbf{g}^T\mathbf{y}(\theta)|\mathbf{P}\mathbf{y}(\theta) = \mathbf{b}(\theta)\},$$

and we assume that we have already found the maximum feasible solution $\mathbf{y}(0)$ corresponding to the vector $\mathbf{b}(0)$.

We have seen [Eqs. (5.1-27), (5.1-28)] that, in the process of obtaining the maximum feasible solution $\mathbf{y}(0)$, the basic feasible solution $\mathbf{y}^{(\nu)}$ at any step ν of the simplex iteration must satisfy

$$\mathbf{T}^{(\nu)}\mathbf{y}^{(\nu)} = (\mathbf{M}^{(\nu)})^{-1}\,\mathbf{b}(0)$$

where $\mathbf{M}^{(\nu)}$ is the matrix whose columns are the basis vectors for step ν. Suppose that the $\mathbf{T}^{(\nu)}$ and the $\mathbf{M}^{(\nu)}$ for the final step of the iteration are denoted by $\mathbf{T}^{(N)}$ and $\mathbf{M}^{(N)}$, respectively. Then the maximum feasible solution $\mathbf{y}(0)$ must satisfy

$$\mathbf{T}^{(N)}\mathbf{y}(0) = [\mathbf{M}^{(N)}]^{-1}\,\mathbf{b}(0). \tag{5.1-56}$$

The locations of the identity matrix columns in $\mathbf{T}^{(N)}$ give all the information needed to determine the nonzero elements of $\mathbf{y}(0)$ from the elements of $[\mathbf{M}^{(N)}]^{-1}\mathbf{b}(0)$. It is also true that any vector $\mathbf{y}(\theta)$ satisfying the constraint $\mathbf{P}\mathbf{y}(\theta) = \mathbf{b}(\theta)$ must also satisfy

$$\mathbf{T}^{(N)}\,\mathbf{y}(\theta) = [\mathbf{M}^{(N)}]^{-1}\,\mathbf{b}(\theta).$$

Substituting Eq. (5.1-55) into this expression gives

$$\mathbf{T}^{(N)}\,\mathbf{y}(\theta) = [\mathbf{M}^{(N)}]^{-1}\,\mathbf{b}(0) + \theta[\mathbf{M}^{(N)}]^{-1}\,\boldsymbol{\beta} \tag{5.1-57}$$

As long as the value of θ is small enough so that all of the elements of the vector on the right-hand side of the foregoing equation remain positive, the maximum feasible solution $\mathbf{y}(\theta)$ corresponding to that value of θ can be obtained from the elements of the right-hand side and the locations of the identity matrix columns in $\mathbf{T}^{(N)}$. That is, the maximum feasible solution $\mathbf{y}(\theta)$ can be obtained from (5.1-57) in the same manner as the maximum feasible solution $\mathbf{y}(0)$ is obtained from (5.1-56). We know that the $\mathbf{y}(\theta)$ obtained in this manner is a maximum feasible solution because all of the nonzero relative cost coefficients $g_j^{(N)}$ will remain negative no matter how we vary θ. If it happens that none of the elements of the vector $[\mathbf{M}^{(N)}]^{-1}\boldsymbol{\beta}$ are negative, then the right-hand side of (5.1-57) will remain positive no matter how large we take θ. In this case the parametric linear program is essentially solved once we obtain the initial maximum feasible solution $\mathbf{y}(0)$ and invert the corresponding basis matrix $\mathbf{M}^{(N)}$. The maximum feasible solution $\mathbf{y}(\theta)$ corresponding to any other $\mathbf{b}(\theta)$ can then be computed by means of Eq. (5.1-57).

If any of the elements of the vector $[\mathbf{M}^{(N)}]^{-1}\boldsymbol{\beta}$ are negative, however, the situation is not so simple. In such an event, there will be a certain value θ_{max} at which some element of the right-hand side of Eq. (5.1-57) becomes zero, and for all larger values of θ, that same element will be negative. It is clear from (5.1-57) that the value of θ_{max} is given by

$$\theta_{max} = \min_{0 \leq i \leq m} \left\{ \frac{\{[\mathbf{M}^{(N)}]^{-1}\,\mathbf{b}(0)\}_i}{|\{[\mathbf{M}^{(N)}]^{-1}\,\boldsymbol{\beta}\}_i|} \,\middle|\, \{[\mathbf{M}^{(N)}]^{-1}\boldsymbol{\beta}\}_i < 0 \right\}$$

where the notation $\{z\}_i$ means the ith element of the vector z. For values of θ greater than θ_{max}, the elements of $[\mathbf{M}^{(N)}]^{-1}\mathbf{b}(0) + \theta\,[\mathbf{M}^{(N)}]^{-1}\boldsymbol{\beta}$ arranged in accordance with the locations of the identity matrix columns in $\mathbf{T}^{(N)}$ will have an element $y_r(\theta)$ that is negative, and hence $\mathbf{y}(\theta)$ cannot be the maximum feasible solution. Therefore, for values of θ greater than θ_{max}, it is necessary to remove the variable $y_r(\theta)$ from the basis set and to replace it with a new variable $y_s(\theta)$ that was previously a nonbasic variable. Once the value of s is determined, this can be accomplished by applying another step of the simplex iteration using the element $t_{r,s}^{(N)}$ as a pivot and obtaining in the process a new tableau $\mathbf{T}^{(N+1)}$ and a new basis matrix $\mathbf{M}^{(N+1)}$. For values of θ greater than θ_{max}, the feasible solution vectors $\mathbf{y}(\theta)$ must then satisfy

$$\mathbf{T}^{(N+1)}\,\mathbf{y}(\theta) = [\mathbf{M}^{(N+1)}]^{-1}\,\mathbf{b}(0) + \theta[\mathbf{M}^{(N+1)}]^{-1}\,\boldsymbol{\beta}, \qquad (5.1\text{-}58)$$

and the value of s must be chosen in such a way that the maximum feasible solution vectors $\mathbf{y}(\theta)$ can be obtained by rearranging the elements of the right-hand side of the foregoing expression in the order indicated by the locations of the identity matrix columns in the new tableau $\mathbf{T}^{(N+1)}$.

In order to simplify the description of how to pick the pivot column s, we assume that columns of the extended tableau

$$\mathbf{Q}^{(N)} = \begin{bmatrix} \mathbf{T}^{(N)} & \mathbf{b}^{(N)} \\ -(\mathbf{g}^{(N)})^T & \gamma_N \end{bmatrix}$$

have been rearranged in such a way that the first m columns of $\mathbf{T}^{(N)}$ are the identity matrix columns in their natural order; that is, we assume

$$\mathbf{Q}^{(N)} = \left[\begin{array}{cccccccc|c} 1 & 0 & \cdots & 0 & t_{1,m+1}^{(N)} & \cdots & t_{1,m+n}^{(N)} & & b_1^{(N)}(\theta) \\ 0 & 1 & \cdots & 0 & t_{2,m+1}^{(N)} & \cdots & t_{2,m+n}^{(N)} & & b_2^{(N)}(\theta) \\ \vdots & \vdots & & \vdots & \vdots & & \vdots & & \vdots \\ 0 & 0 & \cdots & 1 & t_{m,m+1}^{(N)} & \cdots & t_{m,m+n}^{(N)} & & b_m^{(N)}(\theta) \\ \hline 0 & 0 & \cdots & 0 & -g_{m+1}^{(N)} & \cdots & -g_{m+n}^{(N)} & & \gamma_N(\theta) \end{array}\right],$$

so that, for values of θ less than $\theta_{\max}$, the maximum feasible solution $\mathbf{y}(\theta)$ is given by

$$y_i(\theta) = b_i^{(N)}(\theta), \qquad i = 1, 2, \ldots, m,$$

$$y_i(\theta) = 0, \qquad i = m+1, \ldots, m+n.$$

Now, at $\theta = \theta_{\max}$, we have

$$y_r(\theta_{\max}) = b_r^{(N)}(\theta_{\max}) = 0,$$

and for $\theta > \theta_{\max}$,

$$y_r(\theta) = b_r^{(N)}(\theta) < 0.$$

If $y_r(\theta)$ is replaced by $y_s(\theta)$ as a basic variable, then

$$y_s(\theta) = b_r^{(N+1)}(\theta) = \frac{b_r^{(N)}(\theta)}{t_{r,s}^{(N)}}, \qquad \theta > \theta_{\max}.$$

Since $b_r^{(N)}(\theta)$ is negative for $\theta > \theta_{\max}$, and since $y_s(\theta)$ must be positive for $\theta > \theta_{\max}$, it follows that the pivot element $t_{r,s}^{(N)}$ must be chosen to be negative. If there are no negative elements in the pivot row r, then there are no feasible solutions for values of θ greater than $\theta_{\max}$; and so the parametric linear programming process cannot be continued. If there are negative elements in the pivot row, then the pivot element $t_{r,s}^{(N)}$ must be chosen from among them. In order to determine which of them to choose, we must consider the relative cost vectors $\mathbf{g}^{(N)}$ and $\mathbf{g}^{(N+1)}$. From Eq. (5.1-46) it follows that

$$g_j^{(N+1)} = g_j^{(N)} - \frac{g_s^{(N)}}{t_{r,s}^{(N)}} t_{r,j}^{(N)}, \qquad j = 1, 2, \ldots, m+n.$$

Since $t_{r,s}^{(N)}$ must be negative, this last expression can be written

$$g_j^{(N+1)} = g_j^{(N)} + \frac{g_s^{(N)}}{|t_{r,s}^{(N)}|} t_{r,j}^{(N)}, \qquad j = 1, 2, \ldots, m+n. \tag{5.1-59}$$

Now, in order for the new basis to correspond to a maximum feasible solution, it is necessary that all of the nonzero $g_j^{(N+1)}$ be negative; that is, it is necessary that

$$g_j^{(N+1)} \leqslant 0, \qquad j = 1, 2, \ldots, m+n. \tag{5.1-60}$$

Since all of the $g_j^{(N)}$ are already nonpositive, it follows from (5.1-59) that the foregoing condition is satisfied automatically for all those values of j for which $t_{r,j}^{(N)} \geqslant 0$. Thus, we only need to worry about the $t_{r,j}^{(N)}$ that are negative, and for them we can rewrite Eq. (5.1-59) in the form

$$g_j^{(N+1)} = g_j^{(N)} - \frac{g_s^{(N)}}{|t_{r,s}^{(N)}|} |t_{r,j}^{(N)}|, \qquad t_{r,j}^{(N)} < 0.$$

Therefore, by (5.1-60), we must have

$$g_j^{(N)} - \frac{g_s^{(N)}}{|t_{r,s}^{(N)}|} |t_{r,j}^{(N)}| \leqslant 0$$

or

$$\frac{g_s^{(N)}}{|t_{r,s}^{(N)}|} \geqslant \frac{g_j^{(N)}}{|t_{r,j}^{(N)}|}$$

for all j with $t_{r,j}^{(N)} < 0$. This last condition provides us with the criterion for picking the pivot column s. The value of s must be chosen in such a way that

$$\frac{g_s^{(N)}}{|t_{r,s}^{(N)}|} = \max_{1 \leqslant j \leqslant m+n} \left\{ \frac{g_j^{(N)}}{|t_{r,j}^{(N)}|} \,\middle|\, t_{r,j}^{(N)} < 0 \right\}. \tag{5.1-61}$$

Once the pivot element has been selected and the single step of the simplex iteration has been carried out, the situation is described by Eq. (5.1-58). As long as the elements of the vector on the right-hand side of this equation all remain nonnegative, the maximum feasible solution vector $\mathbf{y}(\theta)$ can be determined from those elements and the arrangement of the identity matrix column in $\mathbf{T}^{(N+1)}$. If, for some new value $\theta = \theta_{max}$, an element of the right-hand side becomes negative, then another change of basis (another simplex iteration) is required, if one is possible; that is, if the pivot row contains at least one negative element. The parametric linear programming process can be continued in this manner until one of two situations is attained: (1) the right-hand side is positive for all values of θ; or (2) some element in the right-hand side becomes negative, but it is impossible to continue because the pivot row does not contain any negative elements.

5.2 The Symmetric Method

The simplex algorithm is by far the most widely used method for solving linear programming problems. A less widely known method that deserves more attention is the symmetric algorithm, which was developed by Rockafeller and Talacko. This method has the advantages that it does not require any slack variables and it does not cycle. If the given linear programming problem is

$$\phi_0 = \max_{\mathbf{x} \geqslant \mathbf{o}} \{\mathbf{c}^T \mathbf{x} | \mathbf{A}\mathbf{x} \leqslant \mathbf{b}\}, \tag{5.2-1}$$

where $\mathbf{A}$ is an $m \times n$ matrix, then the corresponding dual problem is

$$\phi_0 = \min_{\mathbf{u} \geqslant \mathbf{o}} \{\mathbf{b}^T \mathbf{u} | \mathbf{u}^T \mathbf{A} \geqslant \mathbf{c}^T\}. \tag{5.2-2}$$

The method begins with an initial tableau of the form

$$\mathbf{Q}^{(0)} = \begin{bmatrix} \mathbf{A} & \mathbf{b} \\ \mathbf{c}^T & 0 \end{bmatrix} \tag{5.2-3}$$

and at each succeeding step generates a new tableau of the form

$$\mathbf{Q}^{(\nu)} = \begin{bmatrix} \mathbf{A}^{(\nu)} & \mathbf{b}^{(\nu)} \\ (\mathbf{c}^{(\nu)})^T & \phi^{(\nu)} \end{bmatrix}. \tag{5.2-4}$$

This tableau represents both the primal and dual problems, and the algorithm solves both problems simultaneously. The elements of the initial vectors **b** and **c** can be a mixture of positive, negative, and zero values. Each step of the algorithm consists of a choice of a pivot $a_{r,s}^{(\nu)}$ (from the elements of $\mathbf{A}^{(\nu)}$) followed by an elimination procedure that replaces the pivot by its reciprocal, divides every other element in the pivot column by the pivot, divides every other element in the pivot row by the negative of the pivot, and computes all the other elements in the new tableau from the equation

$$q_{i,j}^{(\nu+1)} = q_{i,j}^{(\nu)} - \frac{a_{r,j}^{(\nu)} a_{i,s}^{(\nu)}}{a_{r,s}^{(\nu)}}, \qquad i = 1, 2, \ldots, m+1, i \neq r;$$

$$j = 1, 2, \ldots, n+1, \quad j \neq s.$$

The selection rule for the pivot involves the elements of both $(\mathbf{c}^{(\nu)})^T$ and $\mathbf{b}^{(\nu)}$, and the pivot is always an element of $\mathbf{A}^{(\nu)}$; but during the elimination procedure, the elements of $(\mathbf{c}^{(\nu)})^T$ and $\mathbf{b}^{(\nu)}$ and the element $\phi^{(\nu)}$ are treated just like the other elements in the tableau (i.e., the elements of $\mathbf{A}^{(\nu)}$). When the algorithm has converged, the element $\phi^{(\nu)}$ will be equal to ϕ_0, the vector $\mathbf{b}^{(\nu)}$ will contain the elements of the dual solution $\mathbf{u}$ in a scrambled form, and the vector $\mathbf{c}^{(\nu)}$ will contain the elements of the primal solution $\mathbf{x}$ in a scrambled form. It is easy, during the calculations, to carry along two vectors of indicators (corresponding to the two vectors $\mathbf{b}^{(\nu)}$ and $\mathbf{c}^{(\nu)}$), which contain the information needed to unscramble the solution vectors when the method has converged. A complete description of the algorithm has been given by Talacko in his book *Introduction to Linear Programming and Games of Strategy* [5, Chapter 4]. A Fortran subroutine package that uses the algorithm has been written at Oak Ridge National Lab [6] and has proven to be very effective in solving linear programming problems of the type discussed in Chapter 4.

5.3 Quadratic Programming

In Chapter 4, we showed that the constrained linear estimation problem reduces to a linear programming problem whenever the **b**-vectors are distributed in the region defined by one of the expressions

$$\|\mathbf{b} - \bar{\mathbf{b}}\|_{1,\mathbf{s}} = \|\mathbf{e}\|_{1,\mathbf{s}} = \sum_{i=1}^{m} \frac{|e_i|}{s_i} \leqslant 1,$$

$$\|\mathbf{b} - \bar{\mathbf{b}}\|_{\infty,\mathbf{s}} = \|\mathbf{e}\|_{\infty,\mathbf{s}} = \max_{1 \leqslant i \leqslant m} \left\{\frac{|e_i|}{s_i}\right\} \leqslant 1.$$

But whenever the **b**-vectors are distributed in the ellipsoidal region defined by

$$\|\mathbf{b} - \bar{\mathbf{b}}\|_{2,\mathbf{s}} = \|\mathbf{e}\|_{2,\mathbf{s}} = \left[\sum_{i=1}^{m} \frac{|e_i|^2}{s_i^2}\right]^{1/2} \leqslant 1, \qquad (5.3\text{-}1)$$

we saw that, for a given sample vector **b**, the narrowest possible interval

$$I_{\hat{p}} = [\hat{p}^{\mathrm{lo}}, \hat{p}^{\mathrm{up}}] \qquad (5.3\text{-}2)$$

that can be absolutely guaranteed to contain all of the possible values of

$$\bar{p} = \mathbf{w}^T \bar{\mathbf{x}} \tag{5.3-3}$$

for which $\bar{\mathbf{x}}$ satisfies

$$\mathbf{K}\bar{\mathbf{x}} = \bar{\mathbf{b}} \tag{5.3-4}$$

and

$$\bar{\mathbf{x}} \geqslant \mathbf{0} \tag{5.3-5}$$

is the interval whose end points are given by

$$\hat{p}^{\text{lo}} = \min_{\mathbf{x} \geqslant \mathbf{0}} \{\mathbf{w}^T \mathbf{x} | (\mathbf{K}\mathbf{x} - \hat{\mathbf{b}})^T \mathbf{S}^{-2} (\mathbf{K}\mathbf{x} - \hat{\mathbf{b}}) \leqslant 1\}, \tag{5.3-6a}$$

$$\hat{p}^{\text{up}} = \max_{\mathbf{x} \geqslant \mathbf{0}} \{\mathbf{w}^T \mathbf{x} | (\mathbf{K}\mathbf{x} - \hat{\mathbf{b}})^T \mathbf{S}^{-2} (\mathbf{K}\mathbf{x} - \hat{\mathbf{b}}) \leqslant 1\}, \tag{5.3-6b}$$

where $\mathbf{S}^{-2}$ is the square of the diagonal matrix

$$\mathbf{S}^{-1} = \text{diag}\left(\frac{1}{s_1}, \frac{1}{s_2}, \ldots, \frac{1}{s_m}\right). \tag{5.3-7}$$

Thus, the problem to be solved in this case is that of finding the minimum and maximum values of a linear function of the variables $x_1, x_2, \ldots, x_n$ subject to a quadratic constraint and a nonnegativity constraint. Although it is not immediately obvious, these two problems are quadratic programming problems.

Quadratic programming problems are usually formulated as a constrained minimization or maximization of a quadratic objective function of the form

$$q(x_1, x_2, \ldots, x_n) = q(\mathbf{x}) = \mathbf{c}^T \mathbf{x} \pm \mathbf{x}^T \mathbf{Q}\mathbf{x} \tag{5.3-8}$$

where $\mathbf{c}$ is a given n-vector, $\mathbf{Q}$ is a given $n \times n$ symmetric positive-semidefinite matrix, and the sign is chosen to be + or – depending on whether the problem is formulated as a minimization or a maximization. The variables $x_1, x_2, \ldots, x_n$ are constrained to lie in the region defined by the nonnegativity constraints

$$\mathbf{x} \geqslant \mathbf{0} \tag{5.3-9}$$

and by the m linear constraints

$$\mathbf{A}\mathbf{x} \leqslant \mathbf{b} \tag{5.3-10}$$

where $\mathbf{b}$ is a given m-vector and $\mathbf{A}$ is a given $m \times n$ matrix. If the problem is formulated as a minimization, then the objective function must be *convex* in the given constraint region; that is, if $\mathbf{x}^1$ and $\mathbf{x}^2$ are any two points satisfying the constraint conditions (5.3-9) and (5.3-10), then $q(\mathbf{x})$ must have the property that

$$q[\alpha \mathbf{x}^1 + (1 - \alpha)\mathbf{x}^2] \leqslant \alpha q(\mathbf{x}^1) + (1 - \alpha) q(\mathbf{x}^2) \tag{5.3-11}$$

for all α such that $0 \leqslant \alpha \leqslant 1$. Figure 5.6 illustrates a convex function $q(x)$ of a single variable x constrained to be nonnegative and to satisfy the single linear constraint $x \leqslant b$. It can be shown [7] that if $\mathbf{Q}$ is a positive-semidefinite matrix, then the quadratic form $\mathbf{x}^T\mathbf{Q}\mathbf{x}$ is a convex function of $\mathbf{x}$. A linear function $\mathbf{c}^T\mathbf{x}$ is also a convex function, so that, if $\mathbf{Q}$ is positive-semidefinite, then the sum $\mathbf{c}^T\mathbf{x} + \mathbf{x}^T\mathbf{Q}\mathbf{x}$ is a convex function of $\mathbf{x}$. Thus, whenever the quadratic programming problem is formulated as a minimization, the objective function is taken to be

$$q(\mathbf{x}) = \mathbf{c}^T\mathbf{x} + \mathbf{x}^T\mathbf{Q}\mathbf{x}. \tag{5.3-12a}$$

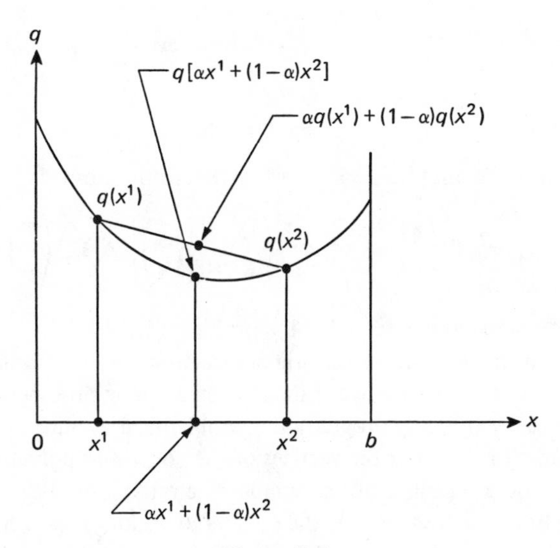

Figure 5.6.

If the problem is formulated as a maximization, then the objective function must be *concave* (a function $f(\mathbf{x})$ is concave in a given region if its negative $-f(\mathbf{x})$ is convex in that region). Since the quadratic form $\mathbf{x}^T\mathbf{Q}\mathbf{x}$ is convex, if $\mathbf{Q}$ is positive semidefinite, the function $-\mathbf{x}^T\mathbf{Q}\mathbf{x}$ is concave. Furthermore, a linear function is also concave (as well as convex), so that the difference $\mathbf{c}^T\mathbf{x} - \mathbf{x}^T\mathbf{Q}\mathbf{x}$ is a concave function of $\mathbf{x}$. Thus, whenever the quadratic programming problem is formulated as a maximization, the objective function is taken to be

$$q(\mathbf{x}) = \mathbf{c}^T\mathbf{x} - \mathbf{x}^T\mathbf{Q}\mathbf{x}. \tag{5.3-12b}$$

In either formulation (minimization or maximization), the matrix $\mathbf{Q}$ is required to be positive semidefinite, primarily because of the present lack of a method for solving the more general problems with an arbitrary symmetric matrix $\mathbf{Q}$.

We can now write as the standard formulations of quadratic programming problems

$$q_0 = \min_{\mathbf{x} \geqslant \mathbf{o}} \{\mathbf{c}^T \mathbf{x} + \mathbf{x}^T \mathbf{Q}\mathbf{x} | \mathbf{A}\mathbf{x} \leqslant \mathbf{b}\} \tag{5.3-13a}$$

and

$$q_0 = \max_{\mathbf{x} \geqslant \mathbf{o}} \{\mathbf{c}^T \mathbf{x} - \mathbf{x}^T \mathbf{Q}\mathbf{x} | \mathbf{A}\mathbf{x} \leqslant \mathbf{b}\}. \tag{5.3-13b}$$

The inequality constraints $\mathbf{Ax} \leqslant \mathbf{b}$ can be replaced by equalities by the same method that was used for the linear programming problem in Section 5.1: by the addition of a vector of slack variables

$$\mathbf{s}^T = (s_1, s_2, \ldots, s_m)$$

satisfying

$$\mathbf{Ax} + \mathbf{s} = \mathbf{b}.$$

This allows the problem (5.3-13a) to be written in the form

$$q_0 = \min_{(\mathbf{x}^T, \mathbf{s}^T) \geqslant \mathbf{o}} \left\{ (\mathbf{c}^T, 0) \begin{pmatrix} \mathbf{x} \\ \mathbf{s} \end{pmatrix} + (\mathbf{x}^T, \mathbf{s}^T) \begin{pmatrix} \mathbf{Q} & 0 \\ 0 & 0 \end{pmatrix} \begin{pmatrix} \mathbf{x} \\ \mathbf{s} \end{pmatrix} \middle| (\mathbf{A}, \mathbf{I}_m) \begin{pmatrix} \mathbf{x} \\ \mathbf{s} \end{pmatrix} = \mathbf{b} \right\}. \tag{5.3-14}$$

A similar statement can be made for the problem (5.3-13b).

The constraint regions for quadratic programming problems are completely similar to the constraint regions for linear programming problems–convex polygons in both cases. In linear programming problems, the solution is guaranteed to occur at one (or more) of the vertices of the constraint polygon. In quadratic programming the situation is not so simple. The minimum value of the objective function might occur anywhere in the constraint region. Figure 5.7 illustrates a problem with $m = 3$ and $n = 2$. The shaded region in the x_1, x_2-plane is the constraint region, and the concentric ovals overlapping this region are contour curves (level surfaces) of the objective function $q(\mathbf{x})$. The minimum value q_0 that $q(\mathbf{x})$ can assume in the constraint region occurs, in this case, on a boundary line between vertices. The solution vector (x_1^0, x_2^0) is the radius vector of the point of tangency between the boundary line and the contour whose equation is

$$q(\mathbf{x}) = q_0.$$

In other problems, the constrained minimum might occur at a vertex of the constraint polygon, as in Fig. 5.8, or in the interior of the constraint polygon, as in Fig. 5.9. In the latter case, the constrained minimum is identical with the absolute minimum value of $q(\mathbf{x})$ when no constraints are present.

The foregoing remarks suggest that quadratic programming problems might be considerably more difficult to solve than linear programming problems. This is indeed true, but nevertheless, several good algorithms exist. Two particularly

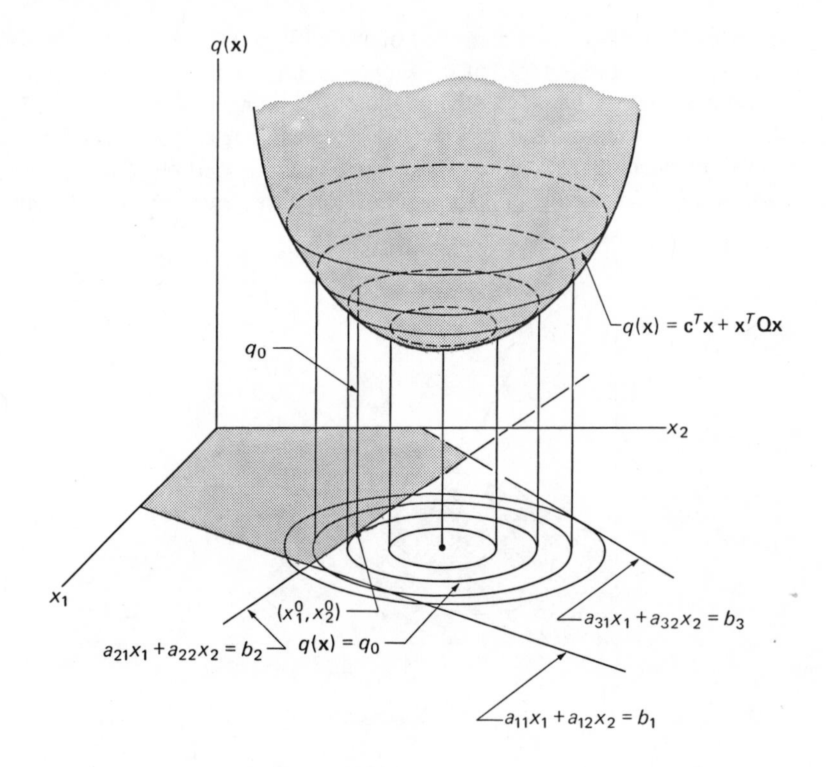

Figure 5.7.

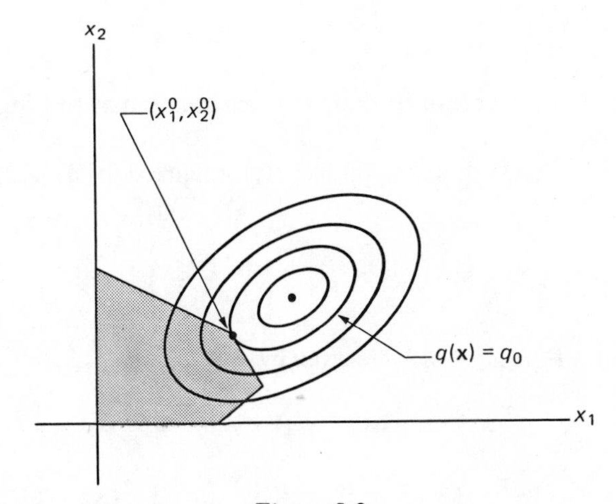

Figure 5.8.

widely used methods are the method of Beale [8] and the method of Wolfe [9], both of which are extensions of the simplex method for linear programming. A good exposition of these and other methods has been given by Kunzi *et al.* [7]. The main disadvantage of most quadratic programming algorithms is the large number of calculations required for convergence to a solution. The simplex algorithm for linear programming requires for convergence roughly the same

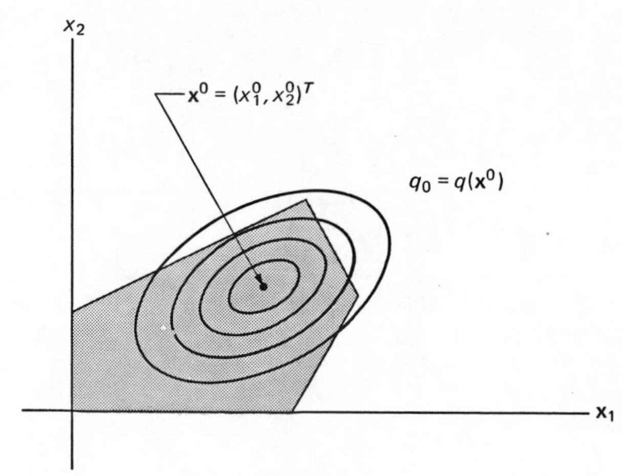

Figure 5.9.

amount of computation as an ordinary matrix inversion (the number of multiplications required is approximately proportional to N^3), but quadratic programming algorithms require much more work and hence much more computer time for convergence.

5.4 Constrained Estimation and Quadratic Programming

When the constrained estimation problem is formulated for the ellipsoidal **b**-vector distribution

$$\|\mathbf{b} - \bar{\mathbf{b}}\|_{2,\mathbf{s}} = \|\mathbf{e}\|_{2,\mathbf{s}} = \left[\sum_{i=1}^{m} \frac{|e_i|^2}{s_i^2}\right]^{1/2} \leqslant 1, \tag{5.4-1}$$

the end points of the interval $I_{\hat{p}}$ are given by

$$\hat{p}^{\text{lo}} = \min_{\mathbf{x} \geqslant \mathbf{0}} \{\mathbf{w}^T \mathbf{x} \mid \|\mathbf{Kx} - \hat{\mathbf{b}}\|_{2,\mathbf{s}}^2 \equiv (\mathbf{Kx} - \hat{\mathbf{b}})^T \mathbf{S}^{-2} (\mathbf{Kx} - \hat{\mathbf{b}}) \leqslant 1\}, \tag{5.4-2}$$

$$\hat{p}^{\text{up}} = \max_{\mathbf{x} \geqslant \mathbf{0}} \{\mathbf{w}^T \mathbf{x} \mid \|\mathbf{Kx} - \hat{\mathbf{b}}\|_{2,\mathbf{s}}^2 \equiv (\mathbf{Kx} - \hat{\mathbf{b}})^T \mathbf{S}^{-2} (\mathbf{Kx} - \hat{\mathbf{b}}) \leqslant 1\}. \tag{5.4-3}$$

In the last section (and in Chapter 4) we stated, but did not prove, that the optimization problems defined by these last two equations are quadratic programming problems. In this section, we will consider the two problems

$$\phi^{\mathrm{lo}} = \min_{\mathbf{x} \geqslant \mathbf{0}} \{\mathbf{w}^T \mathbf{x} | \, \|\mathbf{K}\mathbf{x} - \hat{\mathbf{b}}\|_{2,\mathbf{s}}^2 \leqslant \mu^2\}, \tag{5.4-4}$$

$$\phi^{\mathrm{up}} = \max_{\mathbf{x} \geqslant \mathbf{0}} \{\mathbf{w}^T \mathbf{x} | \, \|\mathbf{K}\mathbf{x} - \hat{\mathbf{b}}\|_{2,\mathbf{s}}^2 \leqslant \mu^2\}, \tag{5.4-5}$$

where μ is a given constant, and show that they can be reduced to the standard forms [Eqs. (5.3-13a, b)] of quadratic programming problems. When $\mu = 1$, these last two problems are identical with problems (5.4-2) and (5.4-3).

For a given sample vector $\hat{\mathbf{b}}$, the quadratic function

$$\|\mathbf{K}\mathbf{x} - \hat{\mathbf{b}}\|_{2,\mathbf{s}}^2 = (\mathbf{K}\mathbf{x} - \hat{\mathbf{b}})^T \mathbf{S}^{-2} (\mathbf{K}\mathbf{x} - \hat{\mathbf{b}}) \tag{5.4-6}$$

defines the residual surface for the least squares problem

$$\|\mathbf{K}\hat{\mathbf{x}} - \hat{\mathbf{b}}\|_{2,\mathbf{s}}^2 = \min_{\mathbf{x}} \|\mathbf{K}\mathbf{x} - \hat{\mathbf{b}}\|_{2,\mathbf{s}}^2. \tag{5.4-7}$$

This surface is shown for a problem with $n = 2$ Fig. 5.10. Figure 5.10a shows a case in which the rank of $\mathbf{K}$ is 2 and Fig. 5.10b shows a case in which the rank of $\mathbf{K}$ is 1. We saw in Section 4.3 that when $\hat{\mathbf{b}}$ is picked from the region defined by Eq. (5.4-1), then the set

$$B_{h,\mathbf{s}}(\hat{b}) = \{\mathbf{b} | \, \|\mathbf{b} - \hat{\mathbf{b}}\|_{2,\mathbf{s}} \leqslant 1\}$$

is guaranteed to contain the vector $\bar{\mathbf{b}}$. This means that every $\mathbf{x}$-vector in the set

$$\{\mathbf{x}(\bar{\mathbf{b}})\} = \{\mathbf{x} | \mathbf{K}\mathbf{x} = \bar{\mathbf{b}}\}$$

satisfies the norm inequality

$$\|\mathbf{K}\mathbf{x} - \hat{\mathbf{b}}\|_{2,\mathbf{s}} \leqslant 1.$$

Therefore, the minimum value defined by Eq. (5.4-7) must be less than or equal to 1; that is,

$$r_0 \equiv \|\mathbf{K}\hat{\mathbf{x}} - \hat{\mathbf{b}}\|_{2,\mathbf{s}}^2 \leqslant 1$$

and the set

$$\{\mathbf{x} | \, \|\mathbf{K}\mathbf{x} - \hat{\mathbf{b}}\|_{2,\mathbf{s}}^2 \leqslant 1\}$$

must always be a nonempty set. This is shown as a shaded region in Fig. 5.10a, b. If the value of the constant μ is chosen to be greater than or equal to 1 (as in Fig. 5.10), then the set

$$\{\mathbf{x} | \, \|\mathbf{K}\mathbf{x} - \hat{\mathbf{b}}\|_{2,\mathbf{s}}^2 \leqslant \mu^2\}$$

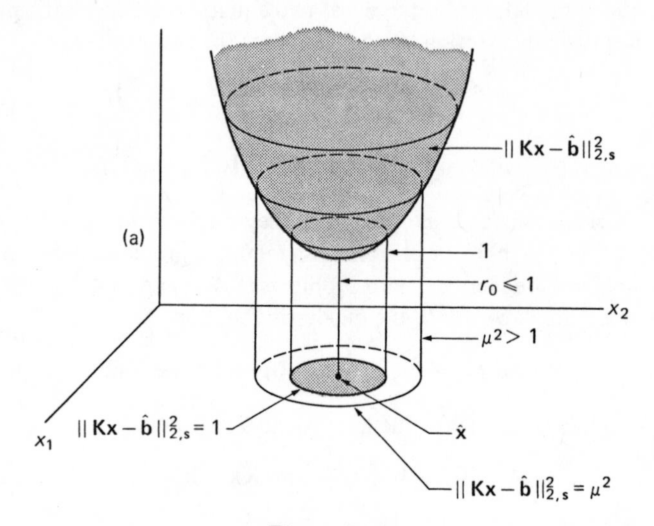

Figure 5.10a.

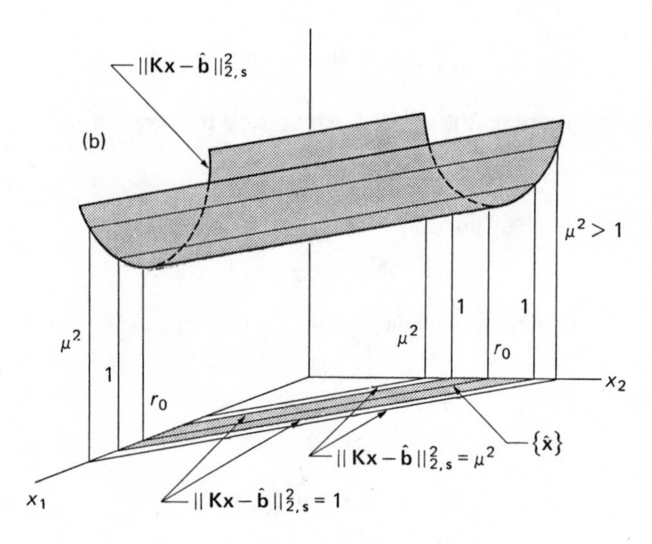

Figure 5.10b.

will also always be a nonempty set, no matter what $\hat{\mathbf{b}}$ is chosen from the region (5.4-1). But if μ is chosen to be less than 1, then for some $\hat{\mathbf{b}}$, this set may be empty. Therefore, as long as we are treating the constrained estimation problem with **b**-vectors distributed according to (5.4-1), we must make the restriction that $\mu \geqslant 1$. The reason for introducting the parameter μ in the first place is that it will be useful later when we treat statistical problems in which we cannot guarantee with certainty that the sample vector $\hat{\mathbf{b}}$ lies in an ellipsoidal region like the one defined by Eq. (5.4-1). The value of μ acts as a scaling factor, and as its value ranges from 0 to ∞, the inequality

$$\|\mathbf{b} - \bar{\mathbf{b}}\|_{2,\mathbf{s}}^2 \leqslant \mu^2$$

defines a nest of concentric ellipsoids, each of which, in the statistical problem, contains a certain amount of the **b** probability density, the amount being proportional to the size of μ^2. Many of the results that we will develop in this and the succeeding chapter can be generalized for use with the statistical problems if the parameter μ is added. For the time being, we will assume that $\mu \geqslant 1$, so that the set $\{\mathbf{x} | \|\mathbf{Kx} - \hat{\mathbf{b}}\|_{2,\mathbf{s}}^2 \leqslant \mu^2\}$ will not fail to exist and will agree to take $\mu = 1$ when we are speaking of constrained estimation problems having a **b** distribution defined by (5.4-1).

In Section 4.4, it was shown that when the Wolfe duality theorem is applied to problems (5.4-2) and (5.4-3), the resulting dual problems are

$$\hat{p}^{\text{lo}} = \max_{\mathbf{u}} \{\mathbf{u}^T \hat{\mathbf{b}} - (\mathbf{u}^T \mathbf{Su})^{1/2} | \mathbf{K}^T \mathbf{u} \leqslant \mathbf{w}\},$$

$$\hat{p}^{\text{up}} = \min_{\mathbf{u}} \{\mathbf{u}^T \hat{\mathbf{b}} + (\mathbf{u}^T \mathbf{Su})^{1/2} | (-\mathbf{K})^T \mathbf{u} \leqslant (-\mathbf{w})\}.$$

In a similar manner, it can be shown that the dual problems for (5.4-4) and (5.4-5) are

$$\phi^{\text{lo}} = \max_{\mathbf{u}} \{\mathbf{u}^T \hat{\mathbf{b}} - \mu(\mathbf{u}^T \mathbf{S}^2 \mathbf{u})^{1/2} | \mathbf{K}^T \mathbf{u} \leqslant \mathbf{w}\}, \tag{5.4-8}$$

$$\phi^{\text{up}} = \min_{\mathbf{u}} \{\mathbf{u}^T \hat{\mathbf{b}} + \mu(\mathbf{u}^T \mathbf{S}^2 \mathbf{u})^{1/2} | (-\mathbf{K})^T \mathbf{u} \leqslant (-\mathbf{w})\}. \tag{5.4-9}$$

These last two problems are very similar in form to the two standard quadratic programming problems (5.3-13a) and (5.3-13b); that is,

$$q_0 = \min_{\mathbf{x} \geqslant \mathbf{0}} \{\mathbf{c}^T \mathbf{x} + \mathbf{x}^T \mathbf{Qx} | \mathbf{Ax} \leqslant \mathbf{b}\}, \tag{5.4-10}$$

$$q_0 = \max_{\mathbf{x} \geqslant \mathbf{0}} \{\mathbf{c}^T \mathbf{x} - \mathbf{x}^T \mathbf{Qx} | \mathbf{Ax} \leqslant \mathbf{b}\}. \tag{5.4-11}$$

But the variables **u** in (5.4-8) and (5.4-9) are not restricted to be nonnegative, and the terms $\pm\mu(\mathbf{u}^T\mathbf{S}^2\mathbf{u})^{1/2}$ are not standard quadratic forms like $\pm\mathbf{x}^T\mathbf{Qx}$. Thus, some other artifice is needed to reduce the constrained estimation problems to the standard quadratic programming forms.

Geometrically, problems (5.4-4) and (5.4-5) consist in finding the two planes of support for the region

$$\{\mathbf{x} |\, \|\mathbf{Kx}-\hat{\mathbf{b}}\|_{2,s}^2 \leq \mu^2\} \cap \{\mathbf{x}|\mathbf{x} \geq \mathbf{0}\}$$

that are orthogonal to the vector **w**. The problem is illustrated for a case with $n = 2$ in Fig. 5.11. The shaded area is the region $\{\mathbf{x}|\, \|\mathbf{Kx}-\hat{\mathbf{b}}\|_{2,s}^2 \leq \mu^2\}$ and the **x**-vectors that solve the problem are the radius vectors of the points of tangency between the support planes and the boundary of this region. In the case shown,

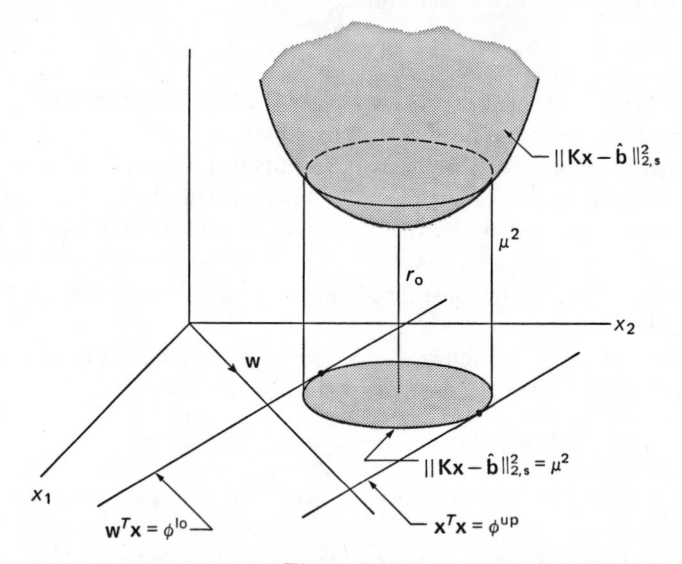

Figure 5.11.

the nonnegativity constraints had no effect on the problem. Figure 5.12 illustrates a whole series of problems, each one corresponding to a different value of μ. For each of these problems, the solution vectors lie on the boundary of the ellipsoidal region defined by the equation

$$\|\mathbf{Kx}-\hat{\mathbf{b}}\|_{2,s}^2 = \mu^2.$$

For the smaller values of μ, the solution points lie on support planes that are tangent to this boundary, but for larger values of μ, the nonnegativity constraints become important, and the solution points lie on the intersection of this boundary with one of the hyperplanes defining the boundary of the region $\{\mathbf{x}|\mathbf{x} \geq 0\}$. If there were no nonnegativity constraints in the problem, then the solution points would always be tangency points, and the problem could be solved by classical Lagrange multiplier techniques. It is the presence of the nonnegativity constraints that transforms the problem from a classical estimation problem into a quadratic programming problem.

Since the solution vectors of problems (5.4-4) and (5.4-5) always lie on the boundary of the region defined by

$$\|\mathbf{Kx} - \hat{\mathbf{b}}\|_{2,\mathbf{s}}^2 \leqslant \mu^2,$$

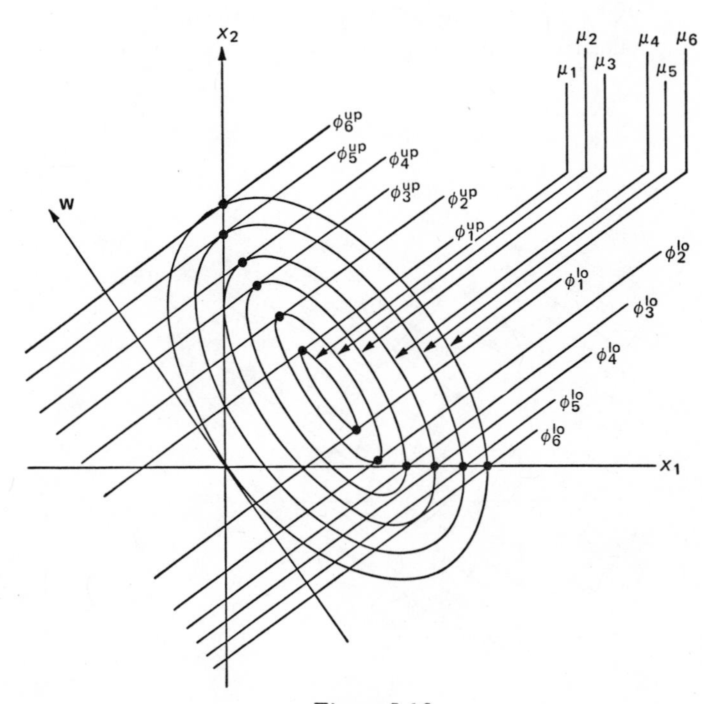

Figure 5.12.

the two problems can also be written

$$\phi^{\mathrm{lo}} = \min_{\mathbf{x} \geqslant \mathbf{0}} \{\mathbf{w}^T \mathbf{x} |\; \|\mathbf{Kx} - \hat{\mathbf{b}}\|_{2,\mathbf{s}}^2 = \mu^2\},$$

$$\phi^{\mathrm{up}} = \max_{\mathbf{x} \geqslant \mathbf{0}} \{\mathbf{w}^T \mathbf{x} |\; \|\mathbf{Kx} - \hat{\mathbf{b}}\|_{2,\mathbf{s}}^2 = \mu^2\}.$$

Furthermore, since the two planes

$$\mathbf{w}^T \mathbf{x} = \phi^{\mathrm{lo}}, \qquad \mathbf{w}^T \mathbf{x} = \phi^{\mathrm{up}},$$

are the support planes for that part of the contour

$$\|\mathbf{Kx} - \hat{\mathbf{b}}\|_{2,\mathbf{s}}^2 = \mu^2$$

which lies in the positive orthant, the solution vectors are the radius vectors of the points at which these two planes make contact with the lowest-level contours that they intersect in the positive orthant. Thus, the two problems can also be written in the inverted form

$$\mu^2 = \min_{\mathbf{x}\geqslant\mathbf{0}} \{\|\mathbf{Kx} - \hat{\mathbf{b}}\|^2_{2,\mathbf{s}} \,|\, \mathbf{w}^T\mathbf{x} = \phi^{\mathrm{lo}}\},$$

$$\mu^2 = \min_{\mathbf{x}\geqslant\mathbf{0}} \{\|\mathbf{Kx} - \hat{\mathbf{b}}\|^2_{2,\mathbf{s}} \,|\, \mathbf{w}^T\mathbf{x} = \phi^{\mathrm{up}}\}.$$

Writing the problems in this inverted form leads to one method for reducing them to the standard quadratic programming problems. The idea is to switch the

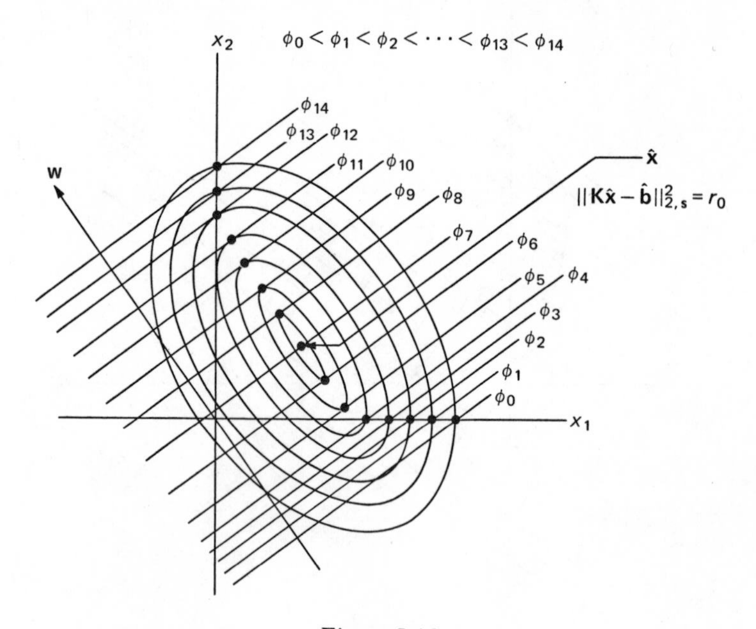

Figure 5.13.

roles of the objective function $\mathbf{w}^T\mathbf{x}$ and the constraint function $\|\mathbf{Kx} - \hat{\mathbf{b}}\|^2_{2,s}$, at the same time introducing a parameter ϕ as the value of the new constraint function. For each value of the parameter ϕ, the new problem can be written

$$L(\phi) = \min_{\mathbf{x}\geqslant\mathbf{0}} \{\|\mathbf{Kx} - \hat{\mathbf{b}}\|^2_{2,\mathbf{s}} \,|\, \mathbf{w}^T\mathbf{x} = \phi\}. \qquad (5.4\text{-}12a)$$

Geometrically, this problem consists in finding, for each value of ϕ, the point at which the hyperplane

$$\mathbf{w}^T\mathbf{x} = \phi$$

makes contact with the lowest-level contour curve of $||\mathbf{Kx}-\hat{\mathbf{b}}||^2_{2,s}$ that it intersects in the positive orthant. The two values ϕ^{lo} and ϕ^{up} are then just the two solutions of the equation

$$L(\phi) = \mu^2. \qquad (5.4\text{-}12b)$$

The problem is illustrated geometrically in Figs. 5.13 and 5.14 for an example with $n = 2$. Figure 5.13 shows a series of contour curves for the new objective function,

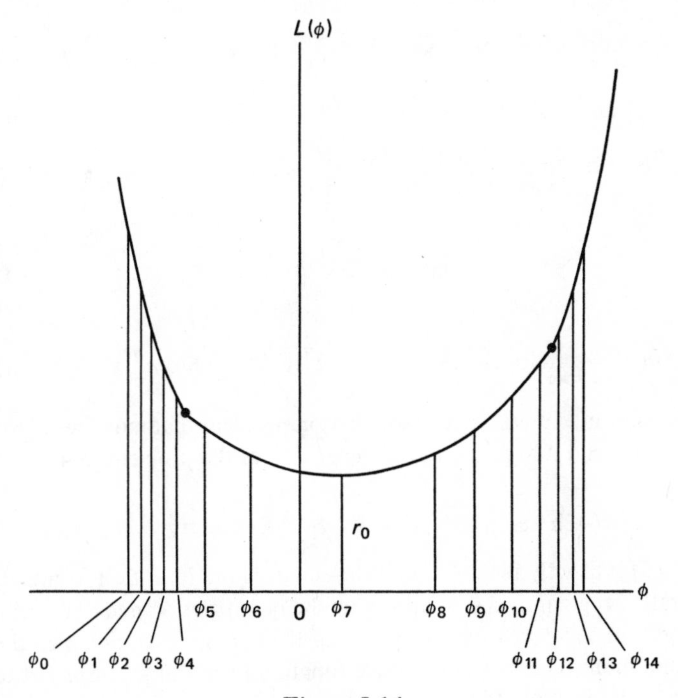

Figure 5.14.

together with 15 different hyperplanes corresponding to 15 different values of the parameter ϕ. The hyperplane defined by

$$\mathbf{w}^T\mathbf{x} = \phi_7$$

passes through the point $\hat{\mathbf{x}}$ at which the objective function $||\mathbf{Kx}-\hat{\mathbf{b}}||^2_{2,s}$ attains its absolute minimum value r_0. For each of the values between ϕ_5 and ϕ_{11}, the solution point is a point of tangency between the corresponding hyperplane and a level curve that is completely contained in the positive orthant. But for some value of ϕ between ϕ_4 and ϕ_5, the nonnegativity constraint on x_1 becomes important, and for some value between ϕ_{11} and ϕ_{12} the nonnegativity constraint on x_2 becomes important. Figure 5.14 is a plot of the minimum value $L(\phi)$ as a function of the parameter ϕ. The curve consists of a parabolic section in the middle (between ϕ_5 and ϕ_{11}) with sharply rising wings joining it on each end. The nodal points at which the wings join the middle section are the points at which the nonnegativity constraints begin to affect the minimum value $L(\phi)$.

For each value of ϕ, the minimization problem (5.4-12a) can be written

$$L(\phi) = \min_{\mathbf{x} \geq \mathbf{0}} \{(\mathbf{Kx} - \hat{\mathbf{b}})^T \mathbf{S}^{-2} (\mathbf{Kx} - \hat{\mathbf{b}}) | \mathbf{w}^T \mathbf{x} = \phi\}$$
$$= \min_{\mathbf{x} \geq \mathbf{0}} \{\mathbf{x}^T \mathbf{K}^T \mathbf{S}^{-2} \mathbf{Kx} - 2\hat{\mathbf{b}}^T \mathbf{S}^{-2} \mathbf{Kx} + \hat{\mathbf{b}}^T \mathbf{S}^{-2} \hat{\mathbf{b}} | \mathbf{w}^T \mathbf{x} = \phi\}.$$

Since $\hat{\mathbf{b}}$ is a given constant sample vector, the problem can also be written

$$L(\phi) = \hat{\mathbf{b}}^T \mathbf{S}^{-2} \hat{\mathbf{b}} + \min_{\mathbf{x} \geq \mathbf{0}} \{(-2\hat{\mathbf{b}}^T \mathbf{S}^{-2} \mathbf{K})\mathbf{x} + \mathbf{x}^T \mathbf{K}^T \mathbf{S}^{-2} \mathbf{Kx} | \mathbf{w}^T \mathbf{x} = \phi\} \quad (5.4\text{-}13)$$

or

$$L(\phi) = \hat{\mathbf{b}}^T \mathbf{S}^{-2} \hat{\mathbf{b}} + q(\phi), \quad (5.4\text{-}14)$$

where

$$q(\phi) = \min_{\mathbf{x} \geq \mathbf{0}} \{(-2\hat{\mathbf{b}}^T \mathbf{S}^{-2} \mathbf{K})\mathbf{x} + \mathbf{x}^T \mathbf{K}^T \mathbf{S}^{-2} \mathbf{Kx} | \mathbf{w}^T \mathbf{x} = \phi\}. \quad (5.4\text{-}15)$$

This last equation expresses the minimization problem that must be solved for each value of ϕ and it has exactly the same form as the problem (5.4-10); that is, if

$$\mathbf{c}^T \equiv (-2\hat{\mathbf{b}}^T \mathbf{S}^{-2} \mathbf{K}), \quad \mathbf{Q} \equiv \mathbf{K}^T \mathbf{S}^{-2} \mathbf{K}, \quad \mathbf{A} \equiv \mathbf{w}^T, \quad \mathbf{b} \equiv \phi,$$

then (5.4-15) is exactly identical with the standard minimization formulation of the quadratic programming problem. Now the quantities ϕ^{lo} and ϕ^{up} defined by (5.4-4) and (5.4-5) are just the two values of the parameter ϕ for which the constrained minimum value of the objective function $||\mathbf{Kx} - \hat{\mathbf{b}}||_{2,s}^2$ turns out to be equal to μ^2; that is, those two values of ϕ that satisfy the equation

$$L(\phi) = \hat{\mathbf{b}}^T \mathbf{S}^{-2} \hat{\mathbf{b}} + q(\phi) = \mu^2 \quad (5.4\text{-}16)$$

(Fig. 5.15 is a graphical illustration of the two solutions of the foregoing equation for the example that was illustrated in Fig. 5.14.) Thus, the problem of determining ϕ^{lo} and ϕ^{up} has been formulated as a problem in *parametric quadratic programming.* At the end of Section 5.1, we discussed an extension of the simplex algorithm that can be used to solve parametric linear programming problems. The method that was outlined can be used to determine how the solution to the linear programming problem varies as the elements of the right-hand-side vector **b** vary continuously in a linear fashion. It is evident that some analogous procedure for quadratic programming is needed in order to solve the problem posed by Eqs. (5.4-15) and (5.4-16). At the end of Section 5.2, we stated, but did not prove, that the standard quadratic programming problem can be solved by an extension of the simplex algorithm. Similarly, the parametric quadratic programming problem posed by (5.4-15) and (5.4-16) can be reduced to a complex problem in parametric linear programming that can be solved by an extension of the method discussed at the end of Section 5.1. We will not discuss the details of the method here, however, because a proper exposition of it is lengthy, complex, and quite difficult. The number of calculations needed to obtain solutions to problems of this sort is also staggering, and it is the hope of the authors that some interested reader will be motivated to discover a more efficient method for solving them.

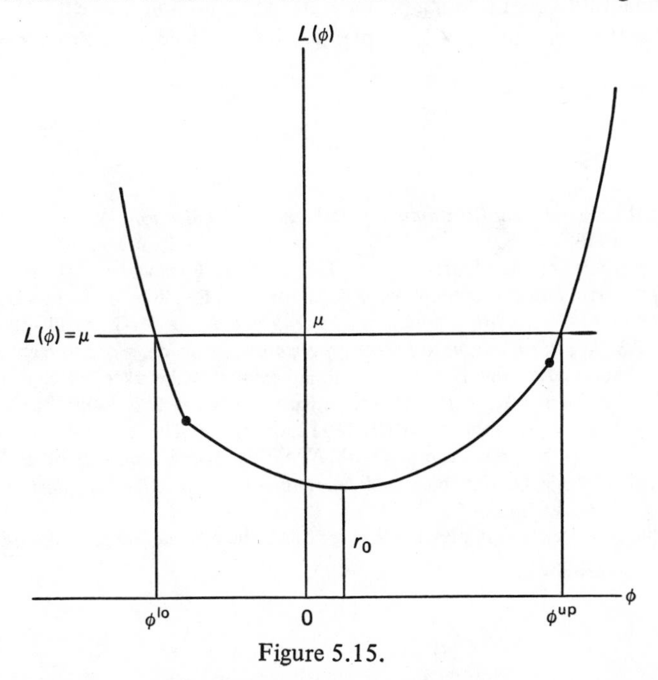

Figure 5.15.

The quadratic programming problem defined by Eq. (5.4-15) requires the formation of the matrix product $\mathbf{K}^T\mathbf{S}^{-2}\mathbf{K}$. The rounding errors that arise in forming this product on a finite-accuracy computer can affect the final solution quite appreciably. Therefore, it is advantageous to avoid forming it. One way to do this is to make the substitution

$$\mathbf{r} = \mathbf{Kx} - \hat{\mathbf{b}}. \tag{5.4-17}$$

We can then write (5.4-12a) in the form

$$\begin{aligned} L(\phi) &= \min_{\mathbf{x} \geqslant \mathbf{0}} \{\|\mathbf{r}\|^2_{2,\mathbf{s}} | \mathbf{r} = \mathbf{Kx} - \hat{\mathbf{b}}, \mathbf{w}^T\mathbf{x} = \phi\} \\ &= \min_{\mathbf{x} \geqslant \mathbf{0}} \{\mathbf{r}^T\mathbf{S}^{-2}\mathbf{r} | \mathbf{Kx} - \mathbf{r} = \hat{\mathbf{b}}, \mathbf{w}^T\mathbf{x} = \phi\}. \end{aligned}$$

The constraints $\mathbf{Kx} - \mathbf{r} = \hat{\mathbf{b}}$ and $\mathbf{w}^T\mathbf{x} = \phi$ can also be written

$$\begin{pmatrix} \mathbf{w}^T & \mathbf{0} \\ \mathbf{K} & -\mathbf{I}_m \end{pmatrix} \begin{pmatrix} \mathbf{x} \\ \mathbf{r} \end{pmatrix} = \begin{pmatrix} \phi \\ \hat{\mathbf{b}} \end{pmatrix},$$

so that the problem can be written

$$L(\phi) = \min_{\mathbf{x} \geqslant \mathbf{0}} \left\{ \mathbf{r}^T \mathbf{S}^{-2} \mathbf{r} \;\middle|\; \begin{pmatrix} \mathbf{w}^T & \mathbf{0} \\ \mathbf{K} & -\mathbf{I} \end{pmatrix} \begin{pmatrix} \mathbf{x} \\ \mathbf{r} \end{pmatrix} = \begin{pmatrix} \phi \\ \hat{\mathbf{b}} \end{pmatrix} \right\}. \qquad (5.4\text{-}18)$$

This quadratic programming problem has more complex constraints than (5.4-15) but the objective function is simpler and it often leads to a simpler implementation.

References

1 Saul I. Gass, *Linear Programming Methods and Applications,* McGraw-Hill, New York, 1958.
2 Ben Noble, *Applied Linear Algebra,* Prentice-Hall, Englewood Cliffs, New Jersey, 1969.
3 G. Hadley, *Linear Programming,* Addison-Wesley, Reading, Massachusetts, 1963.
4 Walter W. Garvin, *Introduction to Linear Programming,* McGraw-Hill, New York, 1960.
5 J. Talacko, *Introduction to Linear Programming and Games of Strategy,* Vol. 1. Marquette Univ. Department of Mathematics, Milwaukee, Wisconsin, 1965.
6 C. Schneeberger, W. R. Burrus, and B. Rust, *A Fortran Subroutine Package for Linear Programming,* ORNL-TM-1250, July 18, 1967.
7 H. P. Kunzi, W. Krelle, and W. Oettli, *Nonlinear Programming,* Blaisdell, New York, 1966.
8 E. M. L. Beale, On quadratic programming, *Naval Res. Logistics Quart.,* 6 (1959), 227–243.
9 Phillip Wolfe, The simplex method for quadratic programming, *Econometrica,* 27 (1959), 382–398.

CHAPTER 6

Applications and Generalizations of the Constrained Estimation Technique

6.1 Constrained Estimation and the Fredholm Integral Equation

In this section, we will develop a method for applying constrained estimation techniques to estimation problems associated with sets of integral equations of the form

$$\int_a^b K_i(s)x(s)\,ds = \hat{y}_i - \hat{e}_i, \qquad i = 1, 2, \ldots, m, \tag{6.1-1}$$

where $x(s)$ is an unknown function, the $K_i(s)$ are known functions, the $\hat{y}_i$ are known values, and the $\hat{e}_i$ are errors that associate the $\hat{y}_i$ with some unknown "true" $\bar{y}_i$; that is,

$$\hat{e}_i = \hat{y}_i - \bar{y}_i. \tag{6.1-2}$$

We will assume that the error vector $\hat{\mathbf{e}}$ is a random sample from some arbitrary distribution in one of the regions defined by

$$\text{I.} \quad \|\mathbf{e}\|_{1,\sigma} \equiv \sum_{i=1}^{m} \frac{|e_i|}{\sigma_i} \leqslant 1, \tag{6.1-3}$$

$$\text{II.} \quad \|\mathbf{e}\|_{2,\sigma} \equiv \left[\sum_{i=1}^{m} \frac{|e_i|^2}{\sigma_i^2}\right]^{1/2} \leqslant 1, \tag{6.1-4}$$

$$\text{III.} \quad \|\mathbf{e}\|_{\infty,\sigma} \equiv \max_{1 \leqslant i \leqslant m} \left[\frac{|e_i|}{\sigma_i}\right] \leqslant 1, \tag{6.1-5}$$

where the σ_i are given positive numbers. The quantities to be estimated have the form

$$\phi = \int_a^b w(s)x(s)\,ds \tag{6.1-6}$$

where $w(s)$ is a known function.

Problems of this type often arise in connection with physical measurements. They were discussed somewhat heuristically in Chapter 3. In this context, the functions $K_i(s)$ are the response functions of some real measuring instrument, and the quantity ϕ can be regarded as the response of an ideal instrument with a response window of transmission $w(s)$. Experimental considerations will usually suggest desirable forms for window functions. The problems under consideration in this section differ from the usual experimental problem in that here the errors are distributed over a bounded set whereas in most experimental situations the errors are assumed to be normally distributed, so that any error is possible with some (perhaps very small) finite probability. The techniques that we will develop here can be easily generalized to this more complex situation; but instead of yielding guaranteed interval estimates, they will give confidence interval estimates.

There are two main classical approaches to solving the estimation problem posed by Eqs. (6.1-1) and (6.1-6). One of these is to assume that the unknown function $x(s)$ can be approximated by an expansion in terms of some linearly independent set of known functions $\{P_k(s)\}$. Of course, if $\{P_k(s)\}$ is a complete orthogonal set, then it is very likely that the unknown $x(s)$ can be expressed as an exact combination of the $P_k(s)$, but the resulting expansion will, in general, have an infinite number of terms; and it is necessary, in order to do actual calculations, to truncate the expansion after some finite number of terms. Therefore, we assume that $x(s)$ has an approximate finite expansion of the form

$$x(s) \cong \sum_{k=1}^{n} g_k P_k(s) \tag{6.1-7}$$

with unknown coefficients g_k. If this expression is substituted into (6.1-1) and the order of integration and summation interchanged, the resulting expression is

$$\sum_{k=1}^{n} g_k \left[\int_a^b K_i(s) P_k(s)\, ds \right] \cong \hat{y}_i - \hat{e}_i, \qquad i = 1, 2, \ldots, m.$$

If **B** is the $m \times n$ matrix whose elements are defined by (and can be calculated from)

$$B_{ik} = \int_a^b K_i(s) P_k(s)\, ds, \tag{6.1-8}$$

then this last expression can be written in vector matrix form as

$$\mathbf{Bg} \cong \hat{\mathbf{y}} - \hat{\mathbf{e}}. \tag{6.1-9}$$

When the expansion (6.1-7) is substituted into Eq. (6.1-6), the result can be written

$$\phi \cong \sum_{k=1}^{n} g_k \left[\int_a^b w(s) P_k(s)\, ds \right].$$

If d_k is the m-vector whose elements are defined by (and can be calculated from)

$$d_k = \int_a^b w(s) P_k(s)\, ds, \tag{6.1-10}$$

then

$$\phi \cong \mathbf{d}^T \mathbf{g}. \tag{6.1-11}$$

Thus, the method seeks to find an approximate estimate of ϕ by solving the classical estimation problem defined by

$$\phi' = \mathbf{d}^T \mathbf{g}' \tag{6.1-11$'$}$$

where

$$\mathbf{B}\mathbf{g}' = \hat{\mathbf{y}} - \hat{\mathbf{e}}. \tag{6.1-9$'$}$$

This problem can be solved by the classical methods discussed in Chapter 2; or if the unknown coefficients g_k can be guaranteed to be nonnegative, by the non-classical methods discussed in Chapter 4. In either case, it is impossible to make any claims about the accuracy of the results, since the unknown function $x(s)$ will in general have a component that is orthogonal to the space spanned by the functions $P_1(s), P_2(s), \ldots, P_n(s)$. This "invisible" component might be very large; and since $x(s)$ is unknown, there is no way to estimate the magnitude of the error introduced by the approximation (6.1-7).

The other classical approach to solving the estimation problem posed by (6.1-1) and (6.1-6) is to expand $K_i(s)$ and $w(s)$ in terms of some linearly independent set of functions $\{P_k(s)\}$. In general, if $\{P_k(s)\}$ is a finite set, then both of the expansions will be approximate. If we write

$$K_i(s) \cong \sum_{k=1}^{n} a_{ik} P_k(s), \qquad i = 1, 2, \ldots, m, \tag{6.1-12}$$

then substitution into (6.1-1) yields

$$\sum_{k=1}^{n} a_{ik} \left[\int_a^b P_k(s) x(s)\, ds \right] \cong \hat{y}_i - \hat{e}_i, \qquad i = 1, 2, \ldots, n.$$

If $\mathbf{v}$ is the unknown n-vector whose elements are defined by

$$v_k = \int_a^b P_k(s) x(s)\, ds, \tag{6.1-13}$$

then the preceding expression can be written in vector matrix form as

$$\mathbf{A}\mathbf{v} \cong \hat{\mathbf{y}} - \hat{\mathbf{e}} \tag{6.1-14}$$

where **A** is the $m \times n$ matrix whose elements are the (known) expansion coefficients a_{ik}. If the approximate expansion for the window function is

$$w(s) \cong \sum_{k=1}^{n} c_k P_k(s), \tag{6.1-15}$$

then substitution into (6.1-6) yields

$$\phi \cong \sum_{k=1}^{n} c_k \left[\int_a^b P_k(s)x(s)\,ds \right],$$

which by (6.1-13) is the same as

$$\phi \cong \mathbf{c}^T \mathbf{v}. \tag{6.1-16}$$

Thus, this method seeks to estimate ϕ by solving the classical estimation problem defined by

$$\phi' = \mathbf{c}^T \mathbf{v}' \tag{6.1-16$'$}$$

where

$$\mathbf{A}\mathbf{v}' = \hat{\mathbf{y}} - \hat{\mathbf{e}}. \tag{6.1-14$'$}$$

Note that this problem has exactly the same form as the problem defined by Eqs. (6.1-9$'$) and (6.1-11$'$), which arose from assuming an expansion for $x(s)$ in terms of the $P_k(s)$. In fact, if the functions $P_k(s)$ are orthonormal on the interval $[a, b]$, and if the approximations used are Fourier-type expansions, then the method of assuming an expansion for $x(s)$ produces exactly the same problem as the one that is obtained by expanding the $K_i(s)$ and $w(s)$ in terms of the $P_k(s)$. The latter method suffers from the same weakness as the former: the presence of components of $x(s)$ that are orthogonal to the space spanned by the $P_k(s)$. There is, in general, no way to estimate the error

$$\mathbf{c}^T \mathbf{v} - \int_a^b w(s)x(s)\,ds,$$

which is induced by the errors in the approximations (6.1-12) and 6.1-15).

There is one way to assure an accurate result when using the classical estimation technique described above. If the expansion functions $P_k(s)$ are chosen to be the response functions $K_i(s)$; that is, if

$$P_k(s) = K_k(s), \qquad k = 1, 2, \ldots, m,$$

then the expansion (6.1-12) is an exact expression (with the matrix **A** equal to the mth-order identity matrix $\mathbf{I}_m$). Furthermore, if the only window functions allowed are those that can be expressed exactly as linear combinations of the response functions; that is,

$$w(s) = \sum_{k=1}^{m} c_k K_k(s),$$

then the solution ϕ' of the classical estimation problem defined by Eqs. (6.1-14′) and (6.1-16′) is exactly equal to the desired ϕ that is defined by (6.1-6). The trouble with this approach, in practical experimental situations, is that the experimenter is very limited in his choice of window functions. At first sight, it might seem that, since the only things known about the function $x(s)$ are the measurements that result from its interaction with the response functions $K_i(s)$, then the only window functions that should be allowed are those that are exact linear combinations of these responses. But in most cases, the situation is not really that bad because the experimenter does have some other (a priori) knowledge about the function $x(s)$. For example, he often knows that the function $x(s)$ is everywhere nonnegative, and this relatively small amount of information makes it possible to give rigorous interval estimates for the interaction of $x(s)$ with window functions that cannot be expressed exactly as a combination of the response functions of the real instrument. In other words, it allows the experimenter to estimate the response to $x(s)$ of an ideal instrument that does not have the shortcomings of the real instrument. Of course, he cannot obtain an exact point estimate, but he can obtain an interval estimate. For the problems we are considering here, the intervals are guaranteed intervals; for problems in which the "measured" values $\hat{y}_i$ are normally distributed, the intervals will be confidence intervals.

The response

$$\phi = \int_a^b w(s)\,x(s)\,ds$$

is said to be *estimable* if the window function $w(s)$ is exactly expressible as a linear combination of the response functions $K_i(s)$; that is, if

$$w(s) = \sum_{i=1}^{m} c_i\,K_i(s)$$

for some constants $c_1, c_2, \ldots, c_m$. This last equation is an expression of the requirement that the estimator for ϕ should be *unbiased.* It is analogous to the unbiasedness condition that was discussed in connection with discrete estimation problems in Section 2.2. A response ϕ that corresponds to a window function that cannot be expressed exactly as a linear combination of the $K_i(s)$ is *inestimable* in the sense that there is no unbiased estimator for it, but if the $x(s)$ is nonnegative everywhere in the interval $[a, b]$, then it is always possible to find a pair of estimable responses ϕ^{lo} and ϕ^{up} that contain the inestimable ϕ in the interval between them. This can be done by finding two sets of coefficients $\{u_1^{\text{lo}}, u_2^{\text{lo}}, \ldots, u_m^{\text{lo}}\}$ and $\{u_1^{\text{up}}, u_2^{\text{up}}, \ldots, u_m^{\text{up}}\}$ that satisfy

$$\sum_{i=1}^{m} u_i^{\text{lo}}\,K_i(s) \leqslant w(s) \leqslant \sum_{i=1}^{m} u_i^{\text{up}}\,K_i(s) \qquad (6.1\text{-}17)$$

where $w(s)$ is the window function for the inestimable response ϕ. The situation is illustrated in Fig. 6.1. The two quantities $\Sigma_{i=1}^{m} u_i^{\text{lo}} K_i(s)$ and $\Sigma_{i=1}^{m} u_i^{\text{up}} K_i(s)$ are taken to be the window functions for the responses ϕ^{lo} and ϕ^{up}. Since $x(s)$ is nonnegative on $[a, b]$, it follows from (6.1-17) that

$$\int_a^b \sum_{i=1}^{m} u_i^{\text{lo}} K_i(s) x(s)\, ds \leqslant \int_a^b w(s) x(s)\, ds \leqslant \int_a^b \sum_{i=1}^{m} u_i^{\text{up}} K_i(s) x(s)\, ds \quad (6.1\text{-}18)$$

or

$$\phi^{\text{lo}} \leqslant \phi \leqslant \phi^{\text{up}}. \quad (6.1\text{-}19)$$

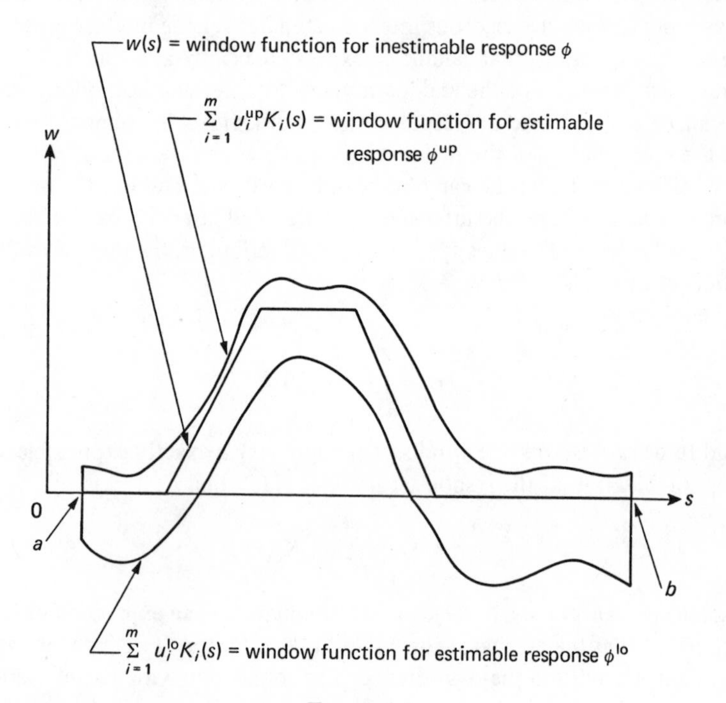

Figure 6.1.

The condition that $x(s)$ be nonnegative on $[a, b]$ is absolutely essential to this method, since otherwise there is no guarantee that the inequalities in expression (6.1-18) follow from those in (6.1-17).

There will, of course, be many sets of coefficients $\{u_1^{\text{lo}}, u_2^{\text{lo}}, \ldots, u_m^{\text{lo}}\}$ and $\{u_1^{\text{up}}, u_2^{\text{up}}, \ldots, u_m^{\text{up}}\}$ that satisfy the inequalities in (6.1-17). Some sets will yield narrower intervals $[\phi^{\text{lo}}, \phi^{\text{up}}]$ than others. Since we want the closest possible estimates for the unknown ϕ, we seek the narrowest possible interval $[\phi^{\text{lo}}, \phi^{\text{up}}]$.

Thus we would like to find that set of coefficients $\{u_1^{up}, u_2^{up}, \ldots, u_m^{up}\}$ which minimizes ϕ^{up} subject to the constraint that

$$w(s) \leqslant \sum_{i=1}^{m} u_i^{up} K_i(s).$$

Similarly, we would also like to find that set $\{u_1^{lo}, u_2^{lo}, \ldots, u_m^{lo}\}$ which maximizes ϕ^{lo} subject to the constraint

$$w(s) \geqslant \sum_{i=1}^{m} u_i^{lo} K_i(s).$$

Although we want to find both the minimum ϕ^{up} and the maximum ϕ^{lo}, we will only discuss the method for finding the minimum ϕ^{up}. The method that will be developed can also be used to find the maximum ϕ^{lo} simply by applying it to the function $-w(s)$ rather than to $w(s)$. We therefore drop the superscript on the u_i^{up} and try to minimize the response

$$\phi^{up} = \int_a^b \left[\sum_{i=1}^{m} u_i K_i(s) \right] x(s)\, ds$$

subject to the *upper biasedness* constraint

$$w(s) \leqslant \sum_{i=1}^{m} u_i K_i(s).$$

That is, we seek to solve the problem

$$\phi^{up} = \min_{(u_1, u_2, \ldots, u_m)} \left\{ \int_a^b \left[\sum_{i=1}^{m} u_i K_i(s) \right] x(s)\, ds \,\middle|\, \sum_{i=1}^{m} u_i K_i(s) \geqslant w(s) \right\}. \qquad (6.1\text{-}20)$$

Now, by Eq. (6.1-1)

$$\begin{aligned} \int_a^b \left[\sum_{i=1}^{m} u_i K_i(s) \right] x(s)\, ds &= \sum_{i=1}^{m} u_i \left[\int_a^b K_i(s) x(s)\, ds \right] \\ &= \sum_{i=1}^{m} u_i (\hat{y}_i - \hat{e}_i) \\ &= u^T (\hat{\mathbf{y}} - \hat{\mathbf{e}}) \end{aligned}$$

where $\hat{\mathbf{e}}$ is an error vector that is known to lie in one of the regions I, II, or III defined by Eqs. (6.1-3)–(6.1-5). Therefore, the problem, (6.1-20) can also be written

$$\phi^{up} = \min_{\mathbf{u}} \left\{ \mathbf{u}^T (\hat{\mathbf{y}} - \hat{\mathbf{e}}) \,\middle|\, \sum_{i=1}^{m} u_i K_i(s) \geqslant w(s) \right\}. \qquad (6.1\text{-}21)$$

We cannot solve this problem as it stands because we do not know the vector $\hat{\mathbf{e}}$. Although $\hat{\mathbf{e}}$ is a fixed vector corresponding to the known vector $\hat{\mathbf{y}}$, all that we really know about it is that it lies somewhere in the region defined by

$$\|\mathbf{e}\|_{h,\sigma} \leqslant 1,$$

where h represents one of the three symbols 1, 2, or ∞.

The error vector $\hat{\mathbf{e}}$ is related to the true but unknown vector $\bar{\mathbf{y}}$ by

$$\bar{\mathbf{y}} = \hat{\mathbf{y}} - \hat{\mathbf{e}}. \tag{6.1-22}$$

What Eq. (6.1-21) says is that we would like ϕ^{up} to be the minimum possible value for $\mathbf{u}^T\bar{\mathbf{y}}$ when $\mathbf{u}$ is constrained by the upper biasedness condition. Since we do not know $\bar{\mathbf{y}}$ exactly, we cannot fulfill this desire. But we do know that

$$\mathbf{u}^T\bar{\mathbf{y}} = \mathbf{u}^T(\hat{\mathbf{y}} - \hat{\mathbf{e}}) \leqslant \max_{\hat{\mathbf{e}}} \{\mathbf{u}^T(\hat{\mathbf{y}} - \hat{\mathbf{e}}) \mid \|\hat{\mathbf{e}}\|_{h,\sigma} \leqslant 1\},$$

so we take ϕ^{up} to be the minimum value of

$$\max_{\hat{\mathbf{e}}} \{\mathbf{u}^T(\hat{\mathbf{y}} - \hat{\mathbf{e}}) \mid \|\hat{\mathbf{e}}\|_{h,\sigma} \leqslant 1\}$$

when $\mathbf{u}$ is constrained by the upper biasedness condition. That is,

$$\phi^{\mathrm{up}} = \min_{\mathbf{u}} \left\{ \max_{\hat{\mathbf{e}}} \{\mathbf{u}^T(\hat{\mathbf{y}} - \hat{\mathbf{e}}) \mid \|\hat{\mathbf{e}}\|_{h,\sigma} \leqslant 1\} \,\middle|\, \sum_{i=1}^{m} u_i K_i(s) \geqslant w(s) \right\}. \tag{6.1-23}$$

The resulting ϕ^{up} will then be an upper bound for ϕ; furthermore, it will be the smallest upper bound that can be obtained with the information that is given.

The expression for the objective function $\max_{\hat{\mathbf{e}}} \{\mathbf{u}^T(\hat{\mathbf{y}} - \hat{\mathbf{e}}) \mid \|\hat{\mathbf{e}}\|_{h,\sigma} \leqslant 1\}$ can be considerably simplified. For simplicity, we assume that $h = 2$ and write

$$\begin{aligned} M_2 &= \max_{\hat{\mathbf{e}}} \{\mathbf{u}^T(\hat{\mathbf{y}} - \hat{\mathbf{e}}) \mid \|\hat{\mathbf{e}}\|_{2,\sigma} \leqslant 1\} \\ &= \max_{\hat{\mathbf{e}}} \{\mathbf{u}^T(\hat{\mathbf{y}} - \hat{\mathbf{e}}) \mid \|\hat{\mathbf{e}}\|_{2,\sigma}^2 \leqslant 1\}. \end{aligned} \tag{6.1-24}$$

If $\mathbf{\Sigma}^{-1}$ is the diagonal matrix

$$\mathbf{\Sigma}^{-1} = \operatorname{diag}\left(\frac{1}{\sigma_1}, \frac{1}{\sigma_2}, \ldots, \frac{1}{\sigma_m}\right),$$

then M_2 can be written

$$M_2 = \max_{\hat{\mathbf{e}}} \{\mathbf{u}^T(\hat{\mathbf{y}} - \hat{\mathbf{e}}) \mid \hat{\mathbf{e}}^T \mathbf{\Sigma}^{-2} \hat{\mathbf{e}} \leqslant 1\}.$$

The problem then is to maximize a linear function of $\hat{\mathbf{e}}$ when $\hat{\mathbf{e}}$ is constrained to lie in a hyperellipsoid. Since there are no other constraints on $\hat{\mathbf{e}}$, it is clear that the maximum will occur somewhere on the boundary of the ellipsoid. Thus we can write

$$M_2 = \max_{\hat{\mathbf{e}}} \{\mathbf{u}^T(\hat{\mathbf{y}} - \hat{\mathbf{e}}) | \hat{\mathbf{e}}^T \boldsymbol{\Sigma}^{-2} \hat{\mathbf{e}} = 1\}. \tag{6.1-25}$$

This problem can be solved by the classical Lagrange multiplier technique. If we write

$$L_2(\hat{\mathbf{e}}) = \mathbf{u}^T(\hat{\mathbf{y}} - \hat{\mathbf{e}}) + \lambda(\hat{\mathbf{e}}^T \boldsymbol{\Sigma}^{-2} \hat{\mathbf{e}} - 1)$$

and carry out the analysis, we find that the Lagrange multiplier λ has the two possible values

$$\lambda^{\pm} = \pm\tfrac{1}{2}(\mathbf{u}^T \boldsymbol{\Sigma}^2 \mathbf{u})^{1/2},$$

and that $\partial L_2/\partial \mathbf{e}$ is equal to zero when $\hat{\mathbf{e}}$ has one of the two values

$$\hat{\mathbf{e}}^{\pm} = \pm (\mathbf{u}^T \boldsymbol{\Sigma}^2 \mathbf{u})^{-1/2} \boldsymbol{\Sigma}^2 \mathbf{u}.$$

Therefore, the maximum value is

$$M_2 = \mathbf{u}^T \hat{\mathbf{y}} + (\mathbf{u}^T \boldsymbol{\Sigma}^2 \mathbf{u})^{1/2}, \tag{6.1-26}$$

which corresponds to

$$\hat{\mathbf{e}} = \hat{\mathbf{e}}^- = -(\mathbf{u}^T \boldsymbol{\Sigma}^2 \mathbf{u})^{-1/2} \boldsymbol{\Sigma}^2 \mathbf{u}.$$

(The value $\hat{\mathbf{e}}^+$ corresponds to a minimum value of L_2.)

If the value in (6.1-26) is substituted for the objective function in (6.1-23), then the resulting expression for ϕ^{up} is

$$\phi^{\mathrm{up}} = \min_{\mathbf{u}} \left\{ \mathbf{u}^T \hat{\mathbf{y}} + (\mathbf{u}^T \boldsymbol{\Sigma}^2 \mathbf{u})^{1/2} \,\middle|\, \sum_{i=1}^{m} u_i K_i(s) \geqslant w(s) \right\},$$

or, using norm notation,

$$\phi^{\mathrm{up}} = \min_{\mathbf{u}} \left\{ \mathbf{u}^T \hat{\mathbf{y}} + \|\mathbf{u}\|_{2,\sigma'} \,\middle|\, \sum_{i=1}^{m} u_i K_i(s) \geqslant w(s) \right\}, \tag{6.1-27}$$

where

$$\|\mathbf{u}\|_{2,\sigma'} \equiv \left[\sum_{i=1}^{m} |u_i|^2 \sigma_i^2 \right]^{1/2}. \tag{6.1-28}$$

In a similar manner, it can be shown that when the error vector $\hat{\mathbf{e}}$ is distributed in the region I, that is,

$$\|\hat{\mathbf{e}}\|_{1,\sigma} = \sum_{i=1}^{m} \frac{|\hat{e}_i|}{\sigma_i} \leqslant 1,$$

then

$$\phi^{\text{up}} = \min_{\mathbf{u}} \left\{ \mathbf{u}^T \hat{\mathbf{y}} + \|\mathbf{u}\|_{\infty,\sigma'} \,\middle|\, \sum_{i=1}^{m} u_i K_i(s) \geqslant w(s) \right\}, \tag{6.1-29}$$

where

$$\|\mathbf{u}\|_{\infty,\sigma'} \equiv \max_{1 \leqslant i \leqslant m} \{|u_i|\sigma_i\}, \tag{6.1-30}$$

and when $\hat{\mathbf{e}}$ is distributed in the region III,

$$\|\hat{\mathbf{e}}\|_{\infty,\sigma} = \max_{1 \leqslant i \leqslant m} \left\{ \frac{|\hat{e}_i|}{\sigma_i} \right\} \leqslant 1,$$

then

$$\phi^{\text{up}} = \min_{\mathbf{u}} \left\{ \mathbf{u}^T \hat{\mathbf{y}} + \|\mathbf{u}\|_{1,\sigma'} \,\middle|\, \sum_{i=1}^{m} u_i K_i(s) \geqslant w(s) \right\} \tag{6.1-31}$$

where

$$\|\mathbf{u}\|_{1,\sigma'} \equiv \sum_{i=1}^{m} |u_i|\sigma_i. \tag{6.1-32}$$

In fact, it can be shown in general that if the error vector $\hat{\mathbf{e}}$ is distributed in any of the regions defined by

$$\|\hat{\mathbf{e}}\|_{h,\sigma} = \left[\sum_{i=1}^{m} \frac{|e_i|^h}{\sigma_i^h} \right]^{1/h} \leqslant 1,$$

where h is any number between 1 and ∞, then the upper bound ϕ^{up} is given by

$$\phi^{\text{up}} = \min_{\mathbf{u}} \left\{ \mathbf{u}^T \hat{\mathbf{y}} + \|\mathbf{u}\|_{h',\sigma'} \,\middle|\, \sum_{i=1}^{m} u_i K_i(s) \geqslant w(s) \right\} \tag{6.1-33}$$

where

$$\|\mathbf{u}\|_{h',\sigma'} \equiv \left[\sum_{i=1}^{m} |u_i|^{h'} \sigma_i^{h'} \right]^{1/h'} \tag{6.1-34}$$

and

$$\frac{1}{h} + \frac{1}{h'} = 1. \tag{6.1-35}$$

The same analysis can be applied to the function $-w(s)$ to obtain expressions for the greatest lower bound ϕ^{lo}; the general result is

$$\phi^{\text{lo}} = \max_{\mathbf{u}} \left\{ \mathbf{u}^T \hat{\mathbf{y}} - \|\mathbf{u}\|_{h',\sigma'} \,\middle|\, \sum_{i=1}^{m} u_i K_i(s) \leqslant w(s) \right\}. \tag{6.1-36}$$

Table 6.1 gives a summary of the problems that must be solved to obtain the interval $[\phi^{\mathrm{lo}}, \phi^{\mathrm{up}}]$ in each of the three cases $h = 1, 2$, and ∞. In each case $[\phi^{\mathrm{lo}}, \phi^{\mathrm{up}}]$ is the narrowest possible interval that can, on the basis of the information given, be guaranteed to contain the true value of ϕ.

The results in Table 6.1 can also be obtained by another method. The quantities ϕ^{lo} and ϕ^{up} can be regarded as the minimum and the maximum value of the functional

$$\phi[x(s)] = \int_a^b w(s)x(s)\,ds$$

when $x(s)$ is constrained to be a solution of one of the sets of integral equations

$$\int_a^b K_i(s)x(s)\,ds = \hat{y}_i - \hat{e}_i, \qquad i = 1, 2, \ldots, n,$$

corresponding to an error vector that is drawn at random from one of the regions $\|\hat{\mathbf{e}}\|_{h,\sigma} \leqslant 1$. Taking into account the nonnegativity constraint on $x(s)$, we can write

$$\phi^{\mathrm{lo}} = \min_{x(s)\geqslant 0} \left\{ \int_a^b w(s)x(s)\,ds \;\middle|\; \left\| \hat{\mathbf{y}} - \left(\int_a^b K_i(s)x(s)\,ds \right) \right\|_{h,\sigma} \leqslant 1 \right\}, \quad (6.1\text{-}37)$$

$$\phi^{\mathrm{up}} = \max_{x(s)\geqslant 0} \left\{ \int_a^b w(s)x(s)\,ds \;\middle|\; \left\| \hat{\mathbf{y}} - \left(\int_a^b K_i(s)x(s)\,ds \right) \right\|_{h,\sigma} \leqslant 1 \right\}, \quad (6.1\text{-}38)$$

where $(\int_a^b K_i(s)x(s)ds)$ denotes the m-vector whose elements are just the quantities $\int_a^b K_i(s)x(s)ds$. In Section 4.4, we saw that if Wolfe's duality theorem is applied to the discrete problems

$$\hat{p}^{\mathrm{lo}} = \min_{\mathbf{x}\geqslant \mathbf{0}} \{\mathbf{w}^T\mathbf{x} \mid \|\mathbf{Kx} - \hat{\mathbf{b}}\|_{h,\sigma} \leqslant 1\},$$

$$\hat{p}^{\mathrm{up}} = \max_{\mathbf{x}\geqslant \mathbf{0}} \{\mathbf{w}^T\mathbf{x} \mid \|\mathbf{Kx} - \hat{\mathbf{b}}\|_{h,\sigma} \leqslant 1\},$$

then the resulting dual problems are

$$\hat{p}^{\mathrm{lo}} = \max_{\mathbf{u}} \{\mathbf{u}^T\hat{\mathbf{b}} - \|\mathbf{u}\|_{h',\sigma'} \mid \mathbf{u}^T\mathbf{K} \leqslant \mathbf{w}^T\},$$

$$\hat{p}^{\mathrm{up}} = \min_{\mathbf{u}} \{\mathbf{u}^T\hat{\mathbf{b}} + \|\mathbf{u}\|_{h',\sigma'} \mid \mathbf{u}^T\mathbf{K} \geqslant \mathbf{w}^T\},$$

where $1/h + 1/h = 1$ and $\boldsymbol{\sigma}'$ is the vector whose elements are the reciprocals of the elements of the vector $\boldsymbol{\sigma}$. In the same manner, the integral $\int_a^b w(s)x(s)ds$ can be regarded as a limit of a sum of the form

$$\int_a^b w(s)x(s)\,ds = \lim_{n\to\infty} \Delta s_n \sum_{i=1}^{n} w_i\,x(s_i)$$

TABLE 6.1

Error vector distribution	Expressions for ϕ^{lo} and ϕ^{up}	Where
$\|\hat{\mathbf{e}}\|_{1,\sigma} = \sum_{i=1}^{m} \frac{\|e_i\|}{\sigma_i} \leqslant 1$	$\phi^{lo} = \max_{\mathbf{u}} \left\{ \mathbf{u}^T \hat{\mathbf{y}} - \|\mathbf{u}\|_{\infty,\sigma'} \middle\| \sum_{i=1}^{m} u_i K_i(s) \leqslant w(s) \right\}$ $\phi^{up} = \min_{\mathbf{u}} \left\{ \mathbf{u}^T \hat{\mathbf{y}} + \|\mathbf{u}\|_{\infty,\sigma'} \middle\| \sum_{i=1}^{m} u_i K_i(s) \geqslant w(s) \right\}$	$\|\mathbf{u}\|_{\infty,\sigma'} = \max_{1 \leqslant i \leqslant m} \{\|u_i\|\sigma_i\}$
$\|\hat{\mathbf{e}}\|_{2,\sigma} = \left[\sum_{i=1}^{m} \frac{\|\hat{e}_i\|^2}{\sigma_i^2} \right]^{1/2} \leqslant 1$	$\phi^{lo} = \max_{\mathbf{u}} \left\{ \mathbf{u}^T \hat{\mathbf{y}} - \|\mathbf{u}\|_{2,\sigma'} \middle\| \sum_{i=1}^{m} u_i K_i(s) \leqslant w(s) \right\}$ $\phi^{up} = \min_{\mathbf{u}} \left\{ \mathbf{u}^T \hat{\mathbf{y}} + \|\mathbf{u}\|_{2,\sigma'} \middle\| \sum_{i=1}^{m} u_i K_i(s) \geqslant w(s) \right\}$	$\|\mathbf{u}\|_{2,\sigma'} = \left[\sum_{i=1}^{m} \|u_i\|^2 \sigma_i^2 \right]^{1/2}$
$\|\hat{\mathbf{e}}\|_{\infty,\sigma} = \max_{1 \leqslant i \leqslant m} \left\{ \frac{\|e_i\|}{\sigma_i} \right\}$	$\phi^{lo} = \max_{\mathbf{u}} \left\{ \mathbf{u}^T \hat{\mathbf{y}} - \|\mathbf{u}\|_{1,\sigma'} \middle\| \sum_{i=1}^{m} u_i K_i(s) \leqslant w(s) \right\}$ $\phi^{up} = \min_{\mathbf{u}} \left\{ \mathbf{u}^T \hat{\mathbf{y}} + \|\mathbf{u}\|_{1,\sigma'} \middle\| \sum_{i=1}^{m} u_i K_i(s) \geqslant w(s) \right\}$	$\|\mathbf{u}\|_{1,\sigma'} = \sum_{i=1}^{m} \|u_i\|\sigma_i$

where $\Delta s_n = (b - a)/n$ and each s_i is some point in the corresponding interval $[a + (i-1)\,\Delta s_n,\ a + (i)\,\Delta s_n]$, and the Wolfe duality theorem can be applied to (6.1-37) and (6.1-38). The resulting dual problems are

$$\phi^{\mathrm{lo}} = \max_{\mathbf{u}} \left\{ \mathbf{u}^T \hat{\mathbf{y}} - \|\mathbf{u}\|_{h',\sigma'} \,\middle|\, \sum_{i=1}^{m} u_i K_i(s) \leq w(s) \right\}, \tag{6.1-39}$$

$$\phi^{\mathrm{up}} = \min_{\mathbf{u}} \left\{ \mathbf{u}^T \hat{\mathbf{y}} + \|\mathbf{u}\|_{h',\sigma'} \,\middle|\, \sum_{i=1}^{m} u_i K_i(s) \geq w(s) \right\}, \tag{6.1-40}$$

which are the same as the problems in Table 6.1.

In Eq. (6.1-40), the problem of estimating ϕ^{up} has been formulated as an extremal problem involving the m-vectors $\mathbf{u}$ and $\hat{\mathbf{y}}$. Only the constraint relation

$$\sum_{i=1}^{m} u_i K_i(s) \geq w(s)$$

involves the continuous variable s. Before ϕ^{up} can actually be computed, this constraint must be discretized in such a way that the continuous constraint is guaranteed to be satisfied provided the discrete approximation is satisfied. A simple way to do this is to replace the window function $w(s)$ by an upper, piecewise linear approximation $w(s)^{\mathrm{up}}$ consisting of a finite number of linear segments with end points at $s_1, s_2, \ldots, s_n$, and to replace each of the response functions $K_i(s)$ by a pair of piecewise linear approximations $K_i(s)^{\mathrm{up}}$ and $K_i(s)^{\mathrm{lo}}$, which bracket it from above and below, and which have the same nodal points $s_1, s_2, \ldots, s_n$ as $w(s)^{\mathrm{up}}$. These approximations are illustrated graphically in Fig. 6.2a, b. The upper approximation $w(s)^{\mathrm{up}}$ satisfies

$$w(s) \leq w(s)^{\mathrm{up}}, \tag{6.1-41}$$

and the approximations $K_i(s)^{\mathrm{lo}}$ and $K_i(s)^{\mathrm{up}}$ satisfy

$$K_i(s)^{\mathrm{lo}} \leq K_i(s) \leq K_i(s)^{\mathrm{up}}, \qquad i = 1, 2, \ldots, m, \tag{6.1-42}$$

at every point s in the interval $[s_1, s_n]$. It is now possible to establish an inequality in terms of these linear approximations that needs only to hold at the node points $s_1, s_2, \ldots, s_n$ to establish a fortiori that the upper biasedness condition holds for the original continuous functions at all s. First, observe that

$$\sum_{i=1}^{m} u_i \begin{bmatrix} K_i(s)^{\mathrm{lo}} \\ K_i(s)^{\mathrm{up}} \end{bmatrix} \leq \sum_{i=1}^{m} u_i K_i(s) \tag{6.1-43}$$

where the upper term in brackets is to be summed if $u_i \geq 0$, and the lower term if $u_i < 0$; that is,

$$\sum_{i=1}^{m} u_i \begin{bmatrix} K_i(s)^{\mathrm{lo}} \\ K_i(s)^{\mathrm{up}} \end{bmatrix} \equiv \sum_{u_i \geq 0} u_i K_i(s)^{\mathrm{lo}} + \sum_{u_i < 0} u_i K_i(s)^{\mathrm{up}}.$$

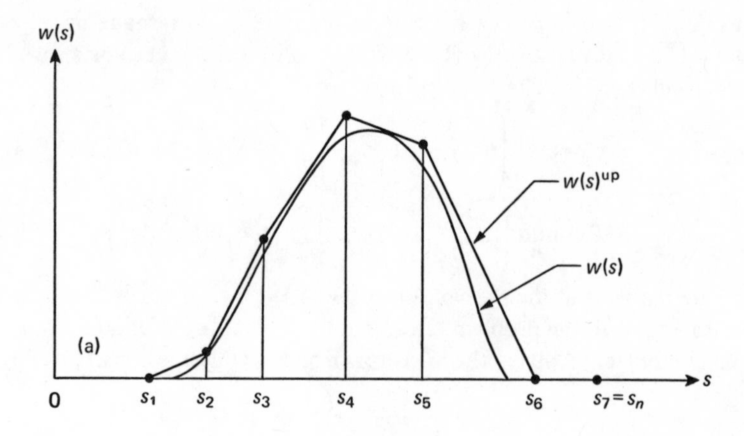

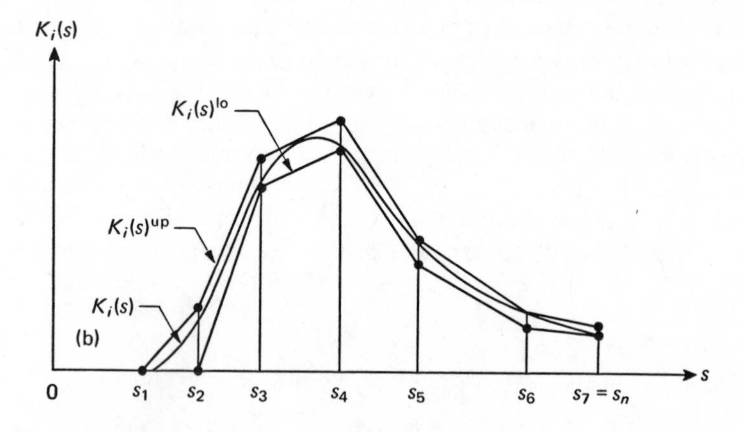

Figure 6.2.

Therefore, if we require the u_i to satisfy

$$\sum_{i=1}^{m} u_i \begin{bmatrix} K_i(s)^{\mathrm{lo}} \\ K_i(s)^{\mathrm{up}} \end{bmatrix} \geqslant w(s)^{\mathrm{up}}, \tag{6.1-44}$$

then the following series of inequalities (illustrated in Fig. 6.3) is implied.

$$\sum_{i=1}^{n} u_i \, K_i(s) \geqslant \sum_{i=1}^{m} u_i \begin{bmatrix} K_i(s)^{\mathrm{lo}} \\ K_i(s)^{\mathrm{up}} \end{bmatrix} \geqslant w(s)^{\mathrm{up}} \geqslant w(s). \tag{6.1-45}$$

Thus, if the u_i satisfy the constraint (6.1-44), then the upper biasedness condition is fulfilled. Note that since the functions $K_i(s)^{lo}$, $K_i(s)^{up}$, and $w(s)^{up}$ are all piecewise linear with the same node points $s_1, s_2, \ldots, s_n$, then it is sufficient to require that the inequality (6.1-44) holds at the node points in order to assure that it holds at all intermediate values of s as well. Thus, the discretized constraint can be written

$$\sum_{i=1}^{m} u_i \begin{bmatrix} K_i(s_j)^{lo} \\ K_i(s_j)^{up} \end{bmatrix} \geqslant w(s_j)^{up}, \qquad j=1,2,\ldots,n. \tag{6.1-46}$$

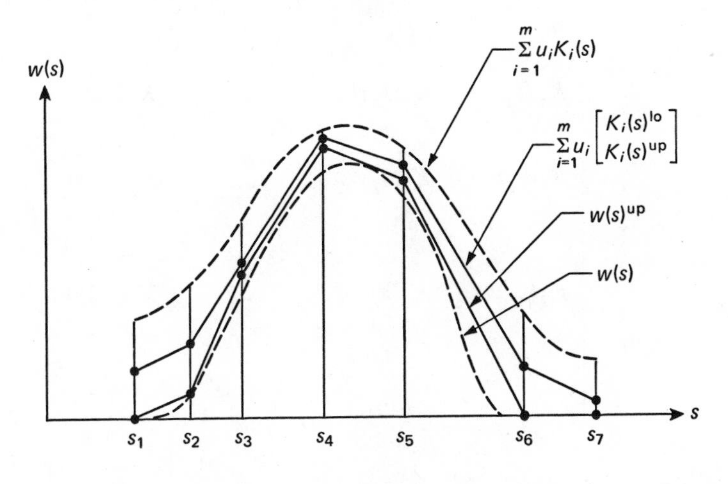

Figure 6.3

The constraint (6.1-46) can be written in a more convenient form if the coefficient vector $\mathbf{u}$ is decomposed into two parts

$$\mathbf{u} = \mathbf{u}^+ - \mathbf{u}^- \tag{6.1-47}$$

where $\mathbf{u}^+$ is a vector whose nonzero components are the positive elements of $\mathbf{u}$, and $\mathbf{u}^-$ is a vector whose nonzero components are the absolute values of the negative elements of $\mathbf{u}$. Thus

$$\mathbf{u}^+ \geqslant \mathbf{0}, \quad \mathbf{u}^- \geqslant \mathbf{0}, \quad (\mathbf{u}^+)^T \mathbf{u}^- = 0, \tag{6.1-48}$$

and the discretized constraint can be written

$$\sum_{i=1}^{m} [u_i^+ K_i(s_j)^{lo} - u_i^- K_i(s_j)^{up}] \geqslant w(s_j)^{up}, j=1,2,\ldots,n. \tag{6.1-49}$$

If $\mathbf{K}^{lo}$ and $\mathbf{K}^{up}$ are the $m \times n$ matrices

$$\mathbf{K}^{lo} = \begin{bmatrix} K_1(s_1)^{lo} & K_1(s_2)^{lo} & \cdots & K_1(s_n)^{lo} \\ K_2(s_1)^{lo} & K_2(s_2)^{lo} & \cdots & K_2(s_n)^{lo} \\ \vdots & \vdots & & \vdots \\ K_m(s_1)^{lo} & K_m(s_2)^{lo} & \cdots & K_m(s_n)^{lo} \end{bmatrix}, \tag{6.1-50}$$

$$\mathbf{K}^{up} = \begin{bmatrix} K_1(s_1)^{up} & K_1(s_2)^{up} & \cdots & K_1(s_n)^{up} \\ K_2(s_1)^{up} & K_2(s_2)^{up} & \cdots & K_2(s_n)^{up} \\ \vdots & \vdots & & \vdots \\ K_m(s_1)^{up} & K_m(s_2)^{up} & \cdots & K_m(s_n)^{up} \end{bmatrix}, \tag{6.1-51}$$

and if $\mathbf{w}^{up}$ is the n-vector

$$\mathbf{w}^{up} = [w(s_1)^{up}, w(s_2)^{up}, \ldots, w(s_n)^{up}]^T, \tag{6.1-52}$$

then the constraint can be written

$$(\mathbf{u}^+)^T K^{lo} - (\mathbf{u}^-)^T \mathbf{K}^{up} \geqslant (\mathbf{w}^{up})^T. \tag{6.1-53}$$

The lower biasedness condition

$$\sum_{i=1}^{m} u_i K_i(s) \leqslant w(s)$$

can be discretized by a method similar to the one used to discretize the upper biasedness condition. If $w(s)^{lo}$ is a piecewise linear lower approximation to $w(s)$ with

$$w(s)^{lo} \leqslant w(s)$$

and with nodes at the points $s_1, s_2, \ldots, s_n$, then the constraint

$$\sum_{i=1}^{m} u_i \begin{bmatrix} K_i(s_j)^{\mathrm{up}} \\ K_i(s_j)^{\mathrm{lo}} \end{bmatrix} \leqslant w(s_j)^{\mathrm{lo}}, \quad j = 1, 2, \ldots, n, \tag{6.1-54}$$

on the u_i guarantees that the lower biasedness condition is satisfied for all values of s in $[s_1, s_n]$. In vector matrix terms this constraint can be written

$$(\mathbf{u}^+)^T \mathbf{K}^{\mathrm{up}} - (\mathbf{u}^-)^T \mathbf{K}^{\mathrm{lo}} \leqslant (\mathbf{w}^{\mathrm{lo}})^T \tag{6.1-55}$$

where $\mathbf{w}^{\mathrm{lo}}$ is the vector

$$\mathbf{w}^{\mathrm{lo}} = [w(s_1)^{\mathrm{lo}}, w(s_2)^{\mathrm{lo}}, \ldots, w(s_n)^{\mathrm{lo}}]^T. \tag{6.1-56}$$

If we replace the lower biasedness condition in (6.1-39) by the constraint (6.1-55) and the upper biasedness condition in (6.1-40) by the constraint (6.1-53), then the resulting problems

$$\phi^{\mathrm{lo}} = \max_{\mathbf{u}} \{\mathbf{u}^T \hat{\mathbf{y}} - \|\mathbf{u}\|_{h',\sigma'} | (\mathbf{u}^+)^T \mathbf{K}^{\mathrm{up}} - (\mathbf{u}^-)^T \mathbf{K}^{\mathrm{lo}} \leqslant (\mathbf{w}^{\mathrm{lo}})^T\}, \tag{6.1-57}$$

$$\phi^{\mathrm{up}} = \min_{\mathbf{u}} \{\mathbf{u}^T \hat{\mathbf{y}} + \|\mathbf{u}\|_{h',\sigma'} | (\mathbf{u}^+)^T \mathbf{K}^{\mathrm{lo}} - (\mathbf{u}^-)^T \mathbf{K}^{\mathrm{up}} \geqslant (\mathbf{w}^{\mathrm{up}})^T\}, \tag{6.1-58}$$

will yield a smaller value for ϕ^{lo} and a larger value for ϕ^{up} than the ones defined by the original problems (6.1-39) and (6.1-40). This widening of the guaranteed interval $[\phi^{\mathrm{lo}}, \phi^{\mathrm{up}}]$ is the price that must be paid in order to get the constraints into a form suitable for computations in a finite numerical method. The amount of widening depends inversely on the number of nodal points $s_1, s_2, \ldots, s_n$ that are used in the discretization. The piecewise linear functions can be made to approximate the continuous functions more and more accurately by using more and more node points, thus reducing the amount of widening of the interval; but the number of calculations that must be done in order to solve the problem also increases as the number of nodes increases. The number of nodes that should be used depends on the amount of widening of $[\phi^{\mathrm{lo}}, \phi^{\mathrm{up}}]$ that one is willing to tolerate in order to reduce the amount of computations. Of course, there is not much point in expending a great deal of effort in order to reduce the amount of widening if the interval defined by (6.1-39) and (6.1-40) is already quite wide because of errors $\hat{\mathbf{e}}$. One possible practical criterion for choosing the number of nodes is to choose n large enough so that the discretization increases the width of the intervals by an amount that is about half the length of the original interval. This number can be determined for a particular class of problems by solving a few test cases with different choices. A sometimes more efficient scheme for discretizing the continuous constraints is to use piecewise parabolic approximations rather than piecewise linear ones. This method has been discussed in a previous work of one of the authors [1].

The problems (6.1-57), (6.1-58) are similar in form to the dual problems that arose from applying the constrained estimation technique to the discrete problems discussed in Chapter 4. When $h' = 1$ or ∞, they can be reduced to linear programming problems and when $h' = 2$, they become quadratic programming problems. The restrictions (6.1-48) on $\mathbf{u}^+$ and $\mathbf{u}^-$ sometimes complicate the solutions. In some problems it is possible to find the solutions by standard linear or quadratic programming methods using only the constraints $\mathbf{u}^+ \geqslant \mathbf{0}$, $\mathbf{u}^- \geqslant \mathbf{0}$ without worrying about the additional constraint $(\mathbf{u}^+)^T \mathbf{u}^- = 0$. In other problems, this procedure leads to unbounded intervals and the additional constraint must be built into the programming algorithm used to solve the problem.

In this section, we have dealt with systems of integral equations of the form

$$\int_a^b K_i(s)\, x(s)\, ds = \hat{y}_i - \hat{e}_i, \qquad i = 1, 2, \ldots, m.$$

The techniques that have been described here can also be used to treat the continuous integral equation

$$\int_a^b K(t, s)\, x(s)\, ds = \hat{y}(t) - \hat{e}(t). \tag{6.1-59}$$

This is done by carrying out the first step of the discretization process described in Section 1.4; that is, by selecting a set of mesh points $t_1, t_2, \ldots, t_m$ and replacing the foregoing equation with the following system of integral equations.

$$\int_a^b K(t_i, s)\, x(s) = \hat{y}(t_i) - \hat{e}(t_i), \qquad i = 1, 2, \ldots, m. \tag{6.1-60}$$

This step introduces many new solution functions $x(s)$ that satisfy the system (6.1-60) but do not satisfy the original equation (6.1-59). Therefore, the intervals $[\phi^{lo}, \phi^{up}]$ will be correspondingly enlarged, but they will still be guaranteed intervals, since any solution of (6.1-59) is also a solution of (6.1-60).

6.2 Generalized Constraints

The nonnegativity condition $x(s) \geqslant 0$ often arises in a natural way in the solution of physically motivated systems of integral equations. Occasionally, the a priori information about $x(s)$ is even more restrictive than the simple nonnegativity condition. For example, $x(s)$ may be known to be monotonically increasing or, perhaps, to be a concave function. In such cases it is often possible to introduce this a priori information into the estimation problems in order to obtain narrower intervals $[\phi^{lo}, \phi^{up}]$. This is done by introducing a new unknown function $z(r)$, defined on some interval $[a', b']$, that satisfies

$$z(r) \geqslant 0 \quad \text{for all} \quad r \in [a', b'] \tag{6.2-1}$$

and is related to the old unknown function by

$$x(s) = \int_{a'}^{b'} R(s,r)\,z(r)\,dr \tag{6.2-2}$$

where $R(s, r)$ is a *regularizing function* that is designed to build in the new restriction on $x(s)$.

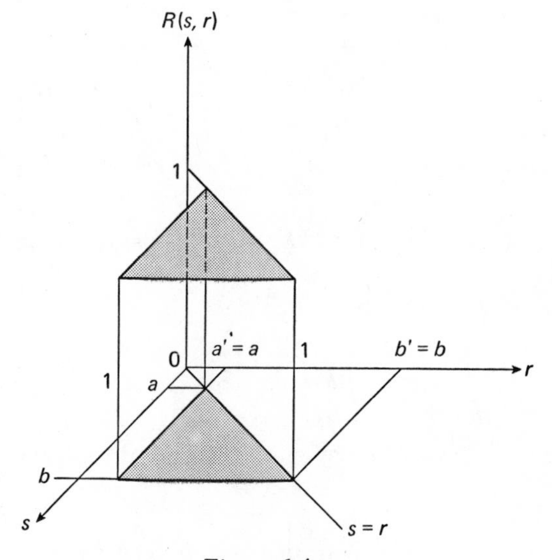

Figure 6.4.

As an example of this kind of transformation, suppose we know that $x(s)$ is nonnegative and monotonically nondecreasing. If $R(s, r)$ is the function (illustrated in Fig. 6.4) defined by

$$R(s,r) = \begin{cases} 1, & \text{if} \quad a \leqslant r, s \leqslant b, \text{ and } r \leqslant s, \\ 0, & \text{otherwise}, \end{cases} \tag{6.2-3}$$

then Eq. (6.2-2) can be written

$$x(s) = \int_a^s 1 \cdot z(r)\,dr + \int_s^b 0 \cdot z(r)\,dr,$$

or simply

$$x(s) = \int_a^s z(r)\,dr. \tag{6.2-4}$$

Clearly, if $z(r)$ is required to be nonnegative on $[a, b]$, then $x(s)$ is required to be nonnegative and monotonically nondecreasing on $[a, b]$.

If Eq. (6.2-2) is substituted into the original system of integral equations (6.1-1), the result is

$$\int_a^b K_i(s)\left[\int_{a'}^{b'} R(s,r)\,z(r)\,dr\right] ds = \hat{y}_i - \hat{e}_i, \qquad i = 1, 2, \ldots, m. \tag{6.2-5}$$

Now

$$\int_a^b K_i(s)\left[\int_{a'}^{b'} R(s,r)\,z(r)\,dr\right] ds = \int_{a'}^{b'} z(r)\left[\int_a^b R(s,r)\,K_i(s)\,ds\right] dr,$$

and if $K_i^t(r)$ is the transformed (or regularized) kernel defined by

$$K_i^t(r) = \int_a^b R(s,r)\,K_i(s)\,ds, \qquad i = 1, 2, \ldots, m, \tag{6.2-6}$$

then the system (6.2-5) can be written

$$\int_{a'}^{b'} K_i^t(r)\,z(r)\,dr = \hat{y}_i - \hat{e}_i, \qquad i = 1, 2, \ldots, m. \tag{6.2-7}$$

Similarly, if (6.2-2) is substituted into Eq. (6.1-6), then the response ϕ can be written

$$\phi = \int_a^b w(s)\left[\int_{a'}^{b'} R(s,r)\,z(r)\,dr\right] ds,$$

and if $w^t(r)$ is the transformed window function defined by

$$w^t(r) = \int_a^b R(s,r)\,w(s)\,ds, \tag{6.2-8}$$

then ϕ can be written

$$\phi = \int_{a'}^{b'} w^t(r)\,z(r)\,dr. \tag{6.2-9}$$

Thus, Eqs. (6.2-7) and (6.2-9) define a transformed problem that (since the only restriction on $z(r)$ is that it be nonnegative) can be treated by the method described in the previous section. The piecewise linear approximations are approximations to the transformed functions $K_i^t(r)$ and $w^t(r)$, which can be obtained from the known functions $K_i(s)$, $w(s)$, and $R(s, r)$ by Eqs. (6.2-6) and (6.2-8). For the example, illustrated in Fig. 6.4, in which $x(s)$ is constrained to be nonnegative and monotonically increasing, these transformed functions are

$$K_i^t(r) = \int_r^b K_i(s)\,ds,$$

$$w^t(r) = \int_r^b w(s)\,ds.$$

The solution to the resulting estimation problem will be an interval $[\phi^{lo}, \phi^{up}]$ that is guaranteed to contain all the possible values of ϕ that are consistent with the given distribution of errors $\hat{e}_i$ and with the restriction that $x(s)$ is a member of the class of functions defined by Eq. (6.2-2).

Regularizing functions can be devised to incorporate all kinds of a priori constraints into the estimation procedure. Another example is to require that $x(s)$ be convex on $[a, b]$. This can be accomplished by requiring that

$$x(s) = \int_a^s \left[\int_a^r v(q)\,dq \right] dr, \qquad v(q) \geqslant 0 \text{ for } q \in [a, b],$$

since then the second derivative of $x(s)$ is

$$\frac{d^2 x}{ds} = v(s),$$

which is required to be nonnegative on $[a, b]$. Furthermore, several regularizing conditions can be applied successively in order to place several simultaneous a priori restrictions on $x(s)$. The regularizing conditions are always applied to the continuous problem, and the final problem that must be discretized always has the same form as Eqs. (6.2-7) and (6.2-9).

6.3. A Second Example of Constrained Estimation

In Section 4.6, we solved for upper and lower bounds to solutions of the Fox-Goodwin problem. In that section, however, the constraints were introduced as matrices multiplying the already discretized form of the kernel and the window vector. In this section, we consider the Bellman problem (see Section 1.5) and apply the monotonicity constraint to the kernel before the discretization step.

The Bellman problem is stated as follows.

$$\int_0^1 (t-s)^2 x(s)\,ds = \frac{t^2}{2} - \frac{2t}{3} + \frac{1}{4}, \qquad 0 \leqslant t \leqslant 1.$$

After the monotonicity constraint was introduced in the manner outlined in the preceding section, the kernel was discretized using $n = 5$ with $s_i = t_i$; the right-hand side was entered to an accuracy of three significant figures, resulting in an error distribution of type III (a linear programming problem). Figure 6.5 shows the upper and lower bounds obtained. These results are no better than those of Replogle *et al.* [2], who solved this problem in the same manner as we solved the Fox-Goodwin problem in Section 4.6. The very large upper bound at $s = 1.0$ arises from the fact that for all values of t_i, the regularized kernel

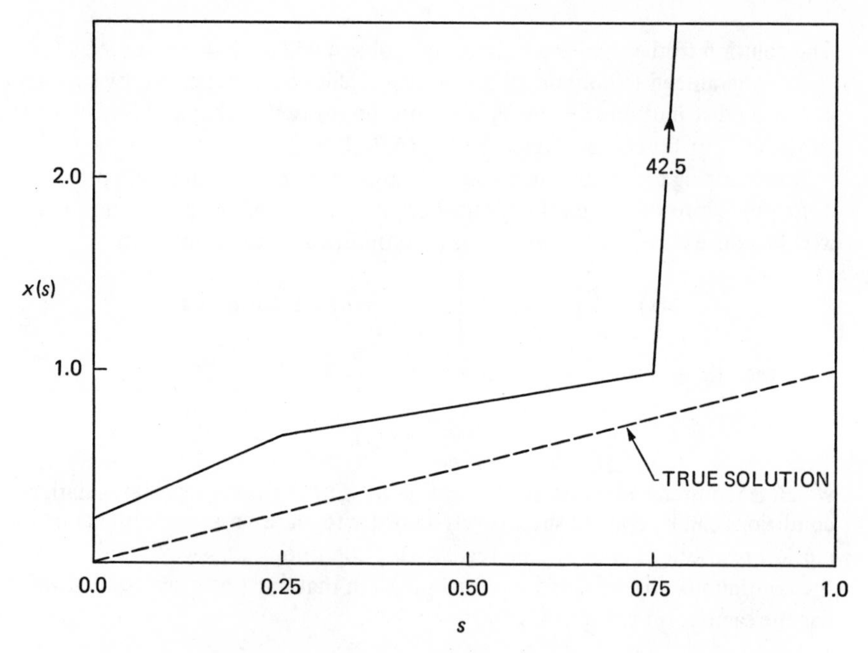

Figure 6.5. Upper and lower (zero line) bounds to Bellman problem.

$K_i^t(r)$ is zero at $r = 1.0$. As a result, the value of x at $s = 1.0$ can be very large and still satisfy the inequalities of the linear programming problem. If the kernel could be infinitely well approximated, the upper bound at $s = 1.0$ would be infinite. However, the errors in the discretization process allow the upper bound to remain finite.

Two methods were used in order to improve the solutions. In the first, a better approximation was made to the window function. The window has the form of a Heaviside step function (e.g., the window for $s = 0.50$ is represented by Fig. 6.6 along with the upper and lower bounds to this window that were used to obtain Figure 6.5). It is obvious that these bounds do not approximate the true function well. Therefore, the kernel and the window were discretized at a series of five *pairs* of points that bracket the five original points a small distance away. This method lessens the area betweeen the true window and its upper and lower bounds by half of what it previously was (see Fig. 6.7). This method did not greatly improve the widths of the intervals (Fig. 6.8).

The errors in fitting upper and lower bounds to the constrained kernel are very large relative to the kernel itself in the region near $s = 1.0$ because the regularized kernel has a value of zero at $r = 1.0$. Therefore, the second method was

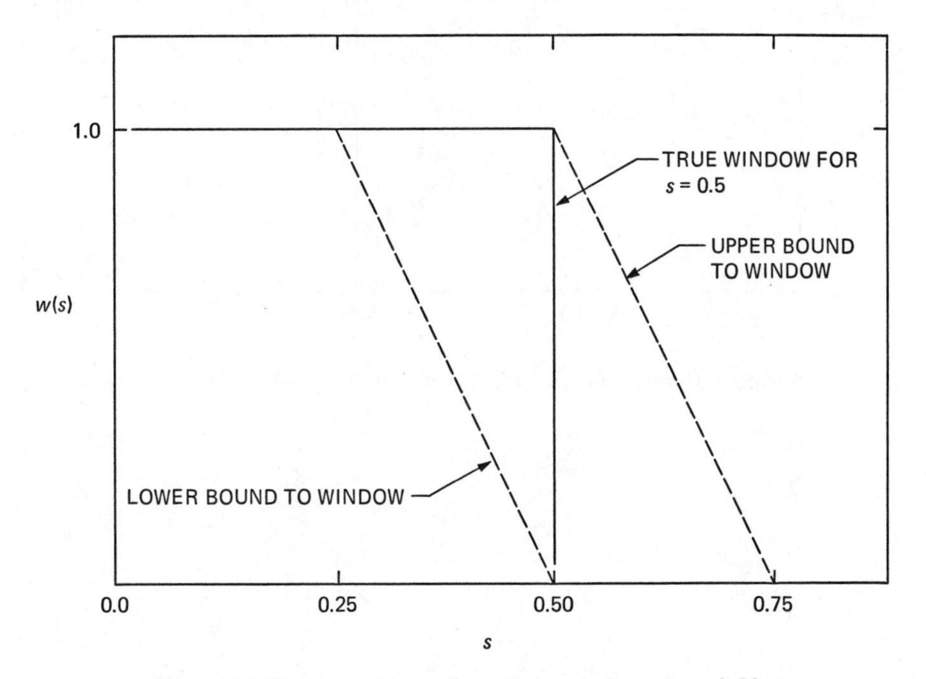

Figure 6.6. Upper and lower bounds to window at $s = 0.50$.

to increase the number of discretization points in the critical region in order to allow the piecewise linear bounds to more closely approximate the curve. Hopefully, the increased accuracy in one sensitive part of the spectrum would aid the solutions in the other parts also.

The new results are shown in Fig. 6.9. The great improvement comes in the region 0.75–1.0 (which was better approximated) and in the fact that the lower bound to the solution is no longer zero. Hence, we see that the sensitive region affects the rest of the spectrum also.

At the risk of repetition, let us emphasize that any monotonicity constraint added to the continuous kernel will cause the regularized kernel to vanish at the

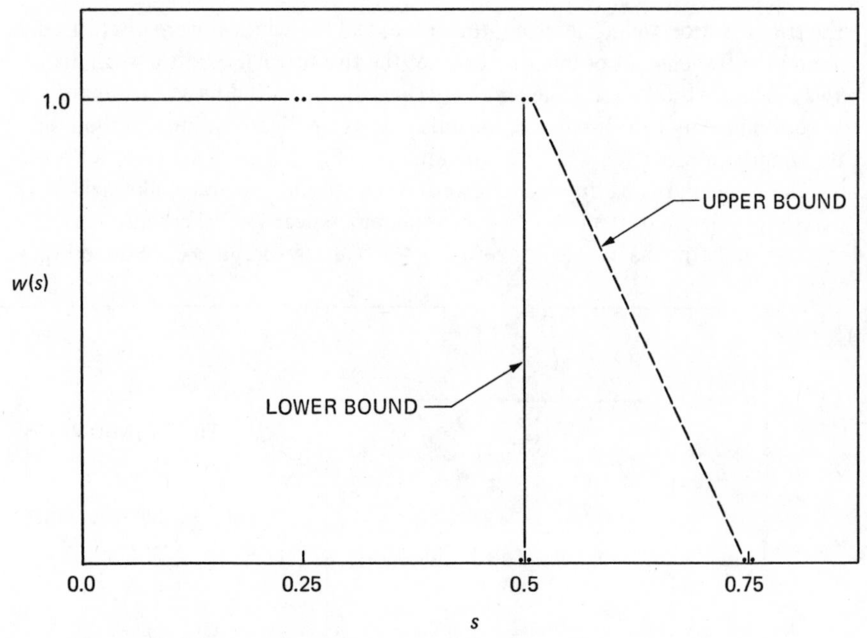

Figure 6.7. Revised upper and lower bounds to window at $s = 0.50$.

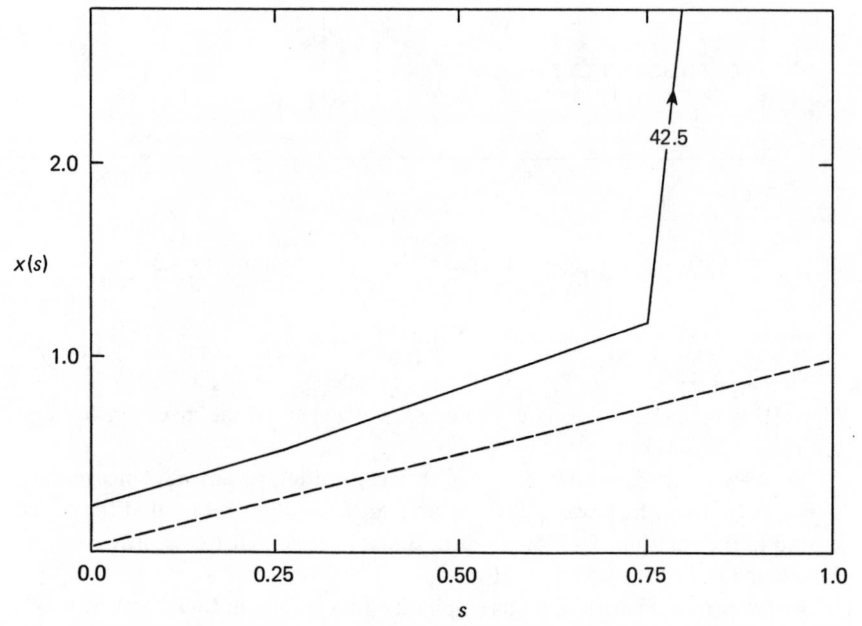

Figure 6.8. Upper and lower (zero line) bounds to Bellman problem.

upper end of its domain (at least if the domain is bounded). Therefore, upper bounds to the solution in this region will be very unstable, perhaps even unbounded.

The discretization using $n = 5$ is very crude, so that the intervals obtained for the solution are wider than they would have been if a larger value of n had been used. The price that would have to be paid for the narrower intervals that would be obtained by increasing n is, of course, more computing time.

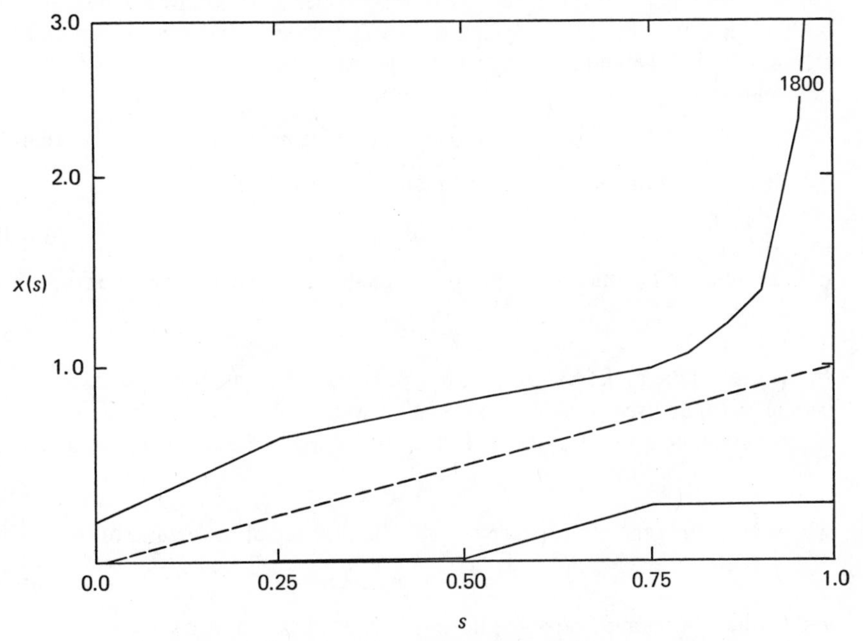

Figure 6.9. Upper and lower bounds to Bellman problem.

It should be noted that the "true solution" shown in Figs. 6.5, 6.8, and 6.9 is really just the particular solution that lies in the subspace spanned by those eigenfunctions of the kernel corresponding to nonzero eigenvalues. There are actually many other solutions of the integral equation, but the intervals obtained in this section are guaranteed to contain all the possible solutions that satisfy the nonnegativity and monotonicity constraints.

6.4 Constrained Estimation of Confidence Intervals

In Chapter 2, we considered estimation problems based on the linear regression model

$$\hat{\mathbf{b}} = \mathbf{A}\bar{\mathbf{x}} + \hat{\boldsymbol{\epsilon}} \tag{6.4-1}$$

where $\mathbf{A}$ is an $m \times n$ matrix of known constants, $\bar{\mathbf{x}}$ is an n-vector of unknown parameters, and $\hat{\mathbf{b}}$ is a stochastic vector of "measured" values that are normally distributed with a mean vector $\mathbf{A}\bar{\mathbf{x}}$ and a positive-definite variance matrix $\mathbf{S}^2$. The probability distribution function for the vector $\hat{\mathbf{b}}$ is

$$g(\hat{\mathbf{b}}) = [(2\pi)^m \det(\mathbf{S}^2)]^{-1/2} \exp[-\tfrac{1}{2}(\hat{\mathbf{b}} - \mathbf{A}\bar{\mathbf{x}})^T \mathbf{S}^{-2}(\hat{\mathbf{b}} - \mathbf{A}\bar{\mathbf{x}})] \quad (6.4\text{-}2)$$

where $\det(\mathbf{S}^2)$ denotes the determinant of the matrix $\mathbf{S}^2$. The vector $\hat{\boldsymbol{\epsilon}}$ is a stochastic vector of errors in the "measured" values $\hat{b}_i$. It is normally distributed with mean vector $\mathbf{0}$ and variance matrix $\mathbf{S}^2$. The classical linear regression model assumes that $m \geqslant n$ and $\operatorname{rank}(\mathbf{A}) = n$, but we will drop both of these restrictions here and let

$$\operatorname{rank}(\mathbf{A}) = r \leqslant \min(m, n). \quad (6.4\text{-}3)$$

The functions that we want to estimate are linear functions

$$\phi = \mathbf{c}^T \bar{\mathbf{x}} \quad (6.4\text{-}4)$$

of the unknown parameters $\bar{x}_i$, and the estimators that we use have the form

$$\hat{\phi} = \mathbf{u}^T \hat{\mathbf{b}}. \quad (6.4\text{-}5)$$

In the case where $\operatorname{rank}(\mathbf{A}) = n \leqslant m$, it is possible to find an unbiased linear estimator for any vector $\mathbf{c}$, but if $\operatorname{rank}(\mathbf{A}) < \min(m, n)$, then it is possible to find an unbiased estimator only if the vector $\mathbf{c}$ satisfies the consistency condition

$$\mathbf{A}^\dagger \mathbf{A} \mathbf{c} = \mathbf{c} \quad (6.4\text{-}6)$$

where $\mathbf{A}^\dagger$ is the generalized inverse of $\mathbf{A}$. The variance of the estimator (6.4-5) is

$$\sigma^2(\hat{\phi}) = \mathbf{u}^T \mathbf{S}^2 \mathbf{u} \quad (6.4\text{-}7)$$

and the best (minimum-variance) linear unbiased estimator is given by

$$\mathbf{u}^T = \mathbf{c}^T (\mathbf{S}^{-1} \mathbf{A})^\dagger \mathbf{S}^{-1} \quad (6.4\text{-}8)$$

where $\mathbf{S}$ is the $m \times m$ matrix defined by

$$\mathbf{S}^2 = \mathbf{S}^T \mathbf{S} = \mathbf{S} \cdot \mathbf{S} \quad (6.4\text{-}9)$$

(such a factorization is always possible for a positive definite $\mathbf{S}^2$). The estimator defined by (6.4-5) and (6.4-8) can also be written

$$\hat{\phi} = \mathbf{c}^T \hat{\mathbf{x}} \quad (6.4\text{-}10)$$

where $\hat{\mathbf{x}}$ is any solution of the least squares problem

$$(\hat{\mathbf{b}} - \mathbf{A}\hat{\mathbf{x}})^T \mathbf{S}^{-2} (\hat{\mathbf{b}} - \mathbf{A}\hat{\mathbf{x}}) = \min_{\mathbf{x}} \{(\hat{\mathbf{b}} - \mathbf{A}\mathbf{x})^T \mathbf{S}^{-2} (\hat{\mathbf{b}} - \mathbf{A}\mathbf{x})\}, \quad (6.4\text{-}11)$$

which can also be written in norm notation as

$$\|\hat{\mathbf{b}} - \mathbf{A}\hat{\mathbf{x}}\|_{\mathbf{S}^2}^2 = \min_{\mathbf{x}} \|\hat{\mathbf{b}} - \mathbf{A}\mathbf{x}\|_{\mathbf{S}^2}^2. \quad (6.4\text{-}12)$$

The estimator $\hat{\phi}$ is a point estimate for ϕ. An interval estimate for ϕ consists of a confidence interval

$$I_\kappa(\phi) = [\hat{\phi} - \kappa\sigma(\hat{\phi}), \hat{\phi} + \kappa\sigma(\hat{\phi})]$$
$$= [\mathbf{u}^T \hat{\mathbf{b}} - \kappa(\mathbf{u}^T \mathbf{S}^2 \mathbf{u})^{1/2}, \mathbf{u}^T \hat{\mathbf{b}} + \kappa(\mathbf{u}^T \mathbf{S}^2 \mathbf{u})^{1/2}], \quad (6.4\text{-}13)$$

which has the property that

$$\Pr\{[\mathbf{u}^T \hat{\mathbf{b}} - \kappa(\mathbf{u}^T \mathbf{S}^2 \mathbf{u})^{1/2}] \leqslant \phi \leqslant [\mathbf{u}^T \hat{\mathbf{b}} + \kappa(\mathbf{u}^T \mathbf{S}^2 \mathbf{u})^{1/2}]\} = \int_{-\kappa}^{\kappa} f(\rho)\, d\rho, \quad (6.4\text{-}14)$$

where $f(\rho)$ is the standard normal distribution function

$$f(\rho) = (2\pi)^{-1/2} \exp[-(\tfrac{1}{2})\rho^2]. \quad (6.4\text{-}15)$$

Combining Eqs. (6.4-7) and (6.4-8) gives

$$\sigma^2(\hat{\phi}) = \mathbf{c}^T (\mathbf{S}^{-1} \mathbf{A})^\dagger [(\mathbf{S}^{-1} \mathbf{A})^\dagger]^T \mathbf{c},$$

which, by Eq. (2.6-9) reduces to

$$\sigma^2(\hat{\phi}) = \mathbf{c}^T (\mathbf{A}^T \mathbf{S}^{-2} \mathbf{A})^\dagger \mathbf{c}. \quad (6.4\text{-}16)$$

Therefore, the confidence interval $I_\kappa(\phi)$ can also be written

$$I_\kappa(\phi) = [\mathbf{c}^T \hat{\mathbf{x}} - \kappa[\mathbf{c}^T(\mathbf{A}^T \mathbf{S}^{-2} \mathbf{A})^\dagger \mathbf{c}]^{1/2}, \mathbf{c}^T \hat{\mathbf{x}} + \kappa[\mathbf{c}^T(\mathbf{A}^T \mathbf{S}^{-2} \mathbf{A})^\dagger \mathbf{c}]^{1/2}]. \quad (6.4\text{-}17)$$

The vectors $\hat{\mathbf{x}}$ defined by Eq. (6.4-11) all satisfy the least squares normal equations

$$(\mathbf{A}^T \mathbf{S}^{-2} \mathbf{A})\hat{\mathbf{x}} = \mathbf{A}^T \mathbf{S}^{-2} \hat{\mathbf{b}}. \quad (6.4\text{-}18)$$

If r_0 denotes the minimum value defined in (6.4-11); that is, if

$$r_0 = (\hat{\mathbf{b}} - \mathbf{A}\hat{\mathbf{x}})^T \mathbf{S}^{-2} (\hat{\mathbf{b}} - \mathbf{A}\hat{\mathbf{x}}) = \hat{\mathbf{b}}^T \mathbf{S}^{-2} \hat{\mathbf{b}} - \hat{\mathbf{x}}^T \mathbf{A}^T \mathbf{S}^{-2} \mathbf{A}\hat{\mathbf{x}}, \quad (6.4\text{-}19)$$

then the equation

$$(\hat{\mathbf{b}} - \mathbf{A}\mathbf{x})^T \mathbf{S}^{-2} (\hat{\mathbf{b}} - \mathbf{A}\mathbf{x}) = r_0 + \kappa^2, \quad (6.4\text{-}20a)$$

which, by Eqs. (6.4-18) and (6.4-19), can also be written

$$(\mathbf{x} - \hat{\mathbf{x}})^T \mathbf{A}^T \mathbf{S}^{-2} \mathbf{A}(\mathbf{x} - \hat{\mathbf{x}}) = \kappa^2, \quad (6.4\text{-}20b)$$

defines the surface of an ellipsoid in x-space (if rank $[\mathbf{A}] < \min[m, n]$, then some of the axes of this ellipsoid will be infinitely long). In Chapter 2, we called this ellipsoid the κ-ellipsoid and saw that if S_- and S_+ are those two support planes of this ellipsoid that are orthogonal to the vector $\mathbf{c}$, then the value of the linear function

$$\phi(\mathbf{x}) = \mathbf{c}^T \mathbf{x}$$

is constant on each of them. The constant values associated with these two support planes are

$$S_-: \quad \mathbf{c}^T\mathbf{x} = \mathbf{c}^T\hat{\mathbf{x}} - \kappa[\mathbf{c}^T(\mathbf{A}^T\mathbf{S}^{-2}\mathbf{A})^\dagger\mathbf{c}]^{1/2}; \tag{6.4-21}$$

$$S_+: \quad \mathbf{c}^T\mathbf{x} = \mathbf{c}^T\hat{\mathbf{x}} + \kappa[\mathbf{c}^T(\mathbf{A}^T\mathbf{S}^{-2}\mathbf{A})^\dagger\mathbf{c}]^{1/2}. \tag{6.4-22}$$

But by Eq. (6.4-17), these are just the end points of the confidence interval $I_\kappa(\phi)$. Thus, the end points of $I_\kappa(\phi)$ are just the minimum and maximum values that the function $\mathbf{c}^T\mathbf{x}$ can assume when $\mathbf{x}$ is constrained to be a point on the surface of the κ-ellipsoid. Therefore, we can write

$$I_\kappa(\phi) = [\phi^{\text{lo}}, \phi^{\text{up}}] \tag{6.4-23}$$

where

$$\phi^{\text{lo}} = \min_{\mathbf{x}} \{\mathbf{c}^T\mathbf{x} | (\hat{\mathbf{b}} - \mathbf{A}\mathbf{x})^T\mathbf{S}^{-2}(\hat{\mathbf{b}} - \mathbf{A}\mathbf{x}) = r_0 + \kappa^2\}, \tag{6.4-24}$$

$$\phi^{\text{up}} = \max_{\mathbf{x}} \{\mathbf{c}^T\mathbf{x} | (\hat{\mathbf{b}} - \mathbf{A}\mathbf{x})^T\mathbf{S}^{-2}(\hat{\mathbf{b}} - \mathbf{A}\mathbf{x}) = r_0 + \kappa^2\}. \tag{6.4-25}$$

Using norm notation, we can also write the interval as

$$I_\kappa(\phi) = \Big[\min_{\mathbf{x}} \{\mathbf{c}^T\mathbf{x} | \, \|\mathbf{A}\mathbf{x} - \hat{\mathbf{b}}\|_{\mathbf{S}^2}^2 = r_0 + \kappa^2\}, \quad \max_{\mathbf{x}} \{\mathbf{c}^T\mathbf{x} | \, \|\mathbf{A}\mathbf{x} - \hat{\mathbf{b}}\|_{\mathbf{S}^2}^2 = r_0 + \kappa^2\}\Big]. \tag{6.4-26}$$

We now assume that we know from a priori considerations that the unknown parameters $\bar{x}_i$ must be nonnegative (as is often the case in physically motivated problems). The two expressions (6.4-24) and (6.4-25) are very similar to the expressions for the end points of the guaranteed interval estimates that were derived in Chapter 4 for the constrained estimation problems in which the vector $\hat{\mathbf{b}}$ was guaranteed with absolute certainty to lie in a finite ellipsoidal region centered at $\bar{\mathbf{b}} = \mathbf{A}\bar{\mathbf{x}}$. Therefore, it is natural to wonder whether we can just add the nonnegativity constraint $\mathbf{x} \geqslant \mathbf{0}$ to the two expressions and still obtain a valid confidence interval. More precisely, the question is whether or not the interval

$$I_\kappa^{\mathbf{c}}(\phi) = \Big[\min_{\mathbf{x}\geqslant\mathbf{0}} \{\mathbf{c}^T\mathbf{x} | \, \|\mathbf{A}\mathbf{x} - \hat{\mathbf{b}}\|_{\mathbf{S}^2}^2 = r_0 + \kappa^2\}, \quad \max_{\mathbf{x}\geqslant\mathbf{0}} \{\mathbf{c}^T\mathbf{x} | \, \|\mathbf{A}\mathbf{x} - \hat{\mathbf{b}}\|_{\mathbf{S}^2}^2 = r_0 + \kappa^2\}\Big] \tag{6.4-27}$$

is a confidence interval that can be guaranteed to contain the true value of ϕ with a probability of

$$\alpha = (2\pi)^{-1/2} \int_{-\kappa}^{\kappa} \exp[-(\tfrac{1}{2})\rho^2]\, d\rho. \tag{6.4-28}$$

Unless the κ-ellipsoid is completely contained in the positive orthant (in which case the constraint $\mathbf{x} \geqslant \mathbf{0}$ adds nothing new to the problem), the interval $I_\kappa^{\mathbf{c}}(\phi)$

in (6.4-27) will, in general, be narrower than the original confidence interval $I_\kappa(\phi)$. The two end points of $I_\kappa^{\mathbf{c}}(\phi)$

$$\phi_{\mathbf{c}}^{\mathbf{lo}} = \min_{\mathbf{x} \geqslant \mathbf{0}} \{\mathbf{c}^T\mathbf{x} \mid \|\mathbf{A}\mathbf{x} - \hat{\mathbf{b}}\|_{\mathbf{S}^2}^2 = r_0 + \kappa^2\}, \tag{6.4-29}$$

$$\phi_{\mathbf{c}}^{\mathbf{up}} = \max_{\mathbf{x} \geqslant \mathbf{0}} \{\mathbf{c}^T\mathbf{x} \mid \|\mathbf{A}\mathbf{x} - \hat{\mathbf{b}}\|_{\mathbf{S}^2}^2 = r_0 + \kappa^2\}, \tag{6.4-30}$$

are just the two constant values assumed by the function $\mathbf{c}^T\mathbf{x}$ on those two support planes of the intersection of the κ-ellipsoid and the positive orthant

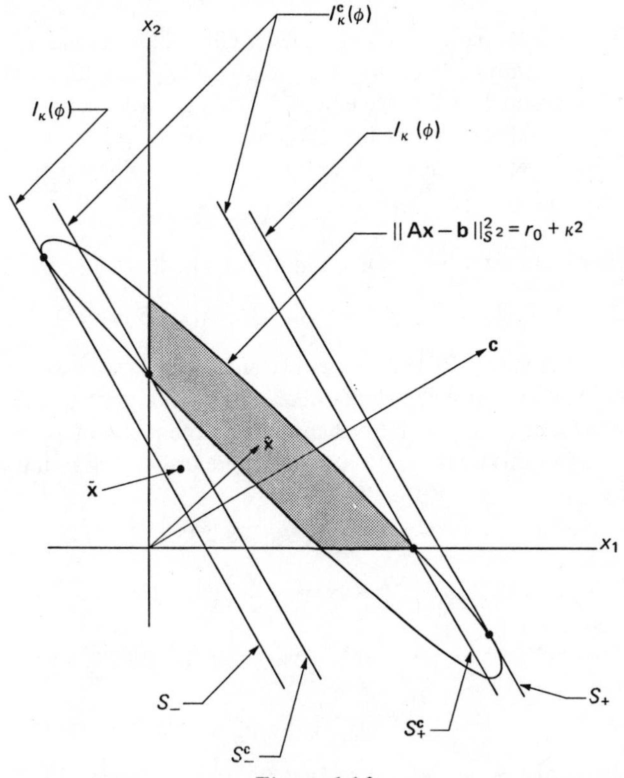

Figure 6.10.

that are orthogonal to the vector $\mathbf{c}$. The equations of these two support planes (denoted by $S_-^{\mathbf{c}}$ and $S_+^{\mathbf{c}}$) are

$$S_-^{\mathbf{c}}: \quad \mathbf{c}^T\mathbf{x} = \phi_{\mathbf{c}}^{\mathbf{lo}}, \tag{6.4-31}$$

$$S_+^{\mathbf{c}}: \quad \mathbf{c}^T\mathbf{x} = \phi_{\mathbf{c}}^{\mathbf{up}}. \tag{6.4-32}$$

The situation is illustrated in Fig. 6.10 for a problem with rank$(\mathbf{A}) = n = 2$. The vector $\bar{\mathbf{x}}$ is guaranteed to lie in the positive orthant with a probability of 100%

and in that region of the positive orthant which is between the two planes S_+ and S_- with a probability of $100\alpha\%$, where α is defined by Eq. (6.4-28). But note that there are many points, like $\tilde{\mathbf{x}}$, that lie in this $100\alpha\%$ region and also in the positive orthant but do not lie between the planes S_-^c and S_+^c. Therefore, we cannot guarantee that $I_\kappa^c(\phi)$ is a $100\alpha\%$ confidence interval for ϕ.

One way to obtain a valid confidence interval for ϕ that takes into account the a priori information that $\bar{\mathbf{x}}$ is nonnegative is to replace the κ-ellipsoid defined by Eq. (6.4-20b), that is,

$$(\mathbf{x}-\hat{\mathbf{x}})^T \mathbf{A}^T \mathbf{S}^{-2} \mathbf{A}(\mathbf{x}-\hat{\mathbf{x}}) = \kappa^2,$$

by a *confidence ellipsoid* for $\bar{\mathbf{x}}$. A confidence ellipsoid is the multidimensional analogue of a confidence interval–or, more precisely, a confidence interval is just a one-dimensional confidence ellipsoid. For a given estimate $\hat{\mathbf{x}}$ of $\bar{\mathbf{x}}$, the $100\alpha\%$ confidence ellipsoid for $\bar{\mathbf{x}}$ is just the ellipsoid in x-space [which we will denote by $E_\gamma(\hat{\mathbf{x}})$] that is centered at the point $\hat{\mathbf{x}}$ and defined by the equation

$$(\mathbf{x}-\hat{\mathbf{x}})^T \mathbf{A}^T \mathbf{S}^{-2} \mathbf{A}(\mathbf{x}-\hat{\mathbf{x}}) = \gamma^2 \tag{6.4-33}$$

where γ^2 is a parameter that is chosen in such a way that

$$\Pr\{\bar{\mathbf{x}} \in E_\gamma(\hat{\mathbf{x}})\} = \Pr\{(\bar{\mathbf{x}}-\hat{\mathbf{x}})^T \mathbf{A}^T \mathbf{S}^{-2} \mathbf{A}(\bar{\mathbf{x}}-\hat{\mathbf{x}}) \leqslant \gamma^2\} = \alpha. \tag{6.4-34}$$

If $\bar{\mathbf{x}}$ is known absolutely to be nonnegative, and is also known to lie in the ellipsoid $E_\gamma(\hat{\mathbf{x}})$ with a probability of $100\alpha\%$, then $\bar{\mathbf{x}}$ is known to lie in the intersection of $E_\gamma(\hat{\mathbf{x}})$ and the positive orthant with a probability of $100\alpha\%$. If the two support planes of this intersection region that are orthogonal to the vector $\mathbf{c}$ are denoted by

$$S_-^E = \{\mathbf{x} | \mathbf{c}^T \mathbf{x} = \phi_E^{\mathrm{lo}}\},$$

$$S_+^E = \{\mathbf{x} | \mathbf{c}^T \mathbf{x} = \phi_E^{\mathrm{up}}\},$$

then $\bar{\mathbf{x}}$ is guaranteed to lie between S_-^E and S_+^E with a probability of $100\alpha\%$; and the interval

$$I_\gamma^E(\phi) = [\phi_E^{\mathrm{lo}}, \phi_E^{\mathrm{up}}] \tag{6.4-35}$$

is a $100\alpha\%$ confidence interval for ϕ. This situation is illustrated in Fig. 6.11 for a problem with rank $(\mathbf{A}) = n = 2$. It is reasonable to expect that the interval $I_\gamma^E(\phi)$ in Eq. (6.4-35) will be wider than the interval $I_\kappa^c(\phi)$ given in Eq. (6.4-27) because the confidence ellipsoid $E_\gamma(\hat{\mathbf{x}})$ will be larger, for a given probability level α, than the κ-ellipsoid corresponding to that value of α. We expect this because the confidence ellipsoid is required to cover the true point $\bar{\mathbf{x}}$ $100\alpha\%$ of the time, but there is no such requirement on the κ-ellipsoid. This widening of the interval is the price that must be paid in order to obtain a valid confidence interval that also takes into account the a priori nonnegativity constraint on $\bar{\mathbf{x}}$.

In order to obtain the relationship between the parameter γ and the probability level α, it is necessary to determine how the quadratic form

$$q(\hat{\mathbf{x}}) = (\bar{\mathbf{x}} - \hat{\mathbf{x}})^T \mathbf{A}^T \mathbf{S}^{-2} \mathbf{A}(\bar{\mathbf{x}} - \hat{\mathbf{x}}) \tag{6.4-36}$$

is distributed. The vector $\hat{\mathbf{b}}$ has a normal distribution with mean vector $\mathbf{A}\bar{\mathbf{x}}$ and variance matrix $\mathbf{S}^2$. We write this in shorthand notation as

$$\hat{\mathbf{b}} \sim N(\mathbf{A}\bar{\mathbf{x}}, \mathbf{S}^2). \tag{6.4-37}$$

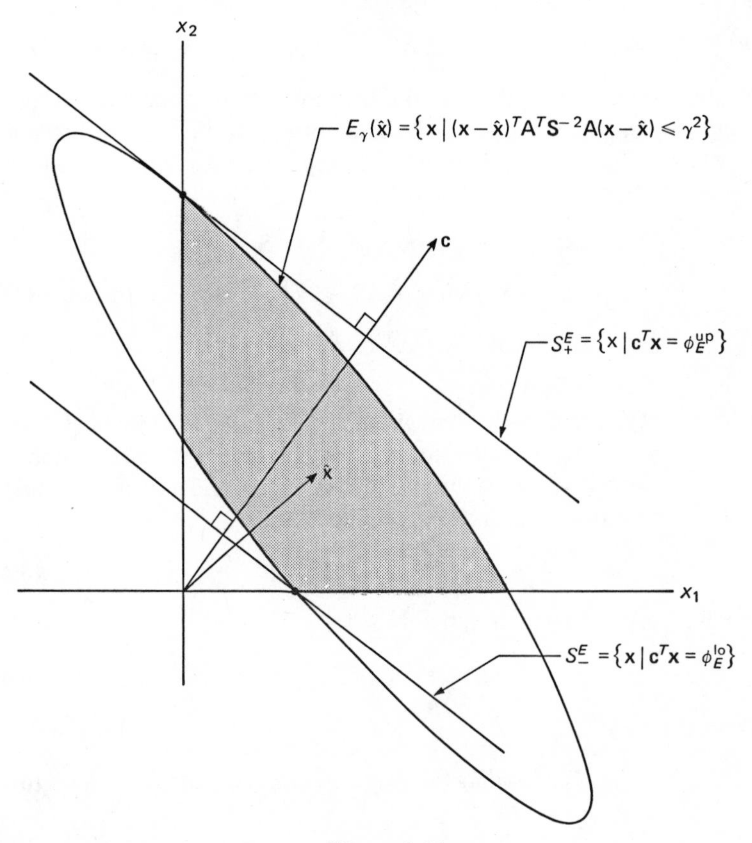

Figure 6.11.

The vector $\hat{\mathbf{x}}$ is given in terms of the vector $\hat{\mathbf{b}}$ by the equation

$$\hat{\mathbf{x}} = (\mathbf{S}^{-1}\mathbf{A})^{\dagger}\mathbf{S}^{-1}\hat{\mathbf{b}},$$

and it can be shown [3, Theorem 2.4.5, p. 26] that $\hat{\mathbf{x}}$ is normally distributed with mean vector $(\mathbf{S}^{-1}\mathbf{A})^{\dagger}\mathbf{S}^{-1}\mathbf{A}\bar{\mathbf{x}}$ and variance matrix

$$\mathbf{\Sigma}_{\hat{\mathbf{x}}} = [(\mathbf{S}^{-1}\mathbf{A})^{\dagger}\mathbf{S}^{-1}]\mathbf{S}^2[(\mathbf{S}^{-1}\mathbf{A})^{\dagger}\mathbf{S}^{-1}]^T = (\mathbf{S}^{-1}\mathbf{A})^{\dagger}[(\mathbf{S}^{-1}\mathbf{A})^{\dagger}]^T = (\mathbf{A}^T\mathbf{S}^{-2}\mathbf{A})^{\dagger}.$$

Thus, we can write

$$\hat{\mathbf{x}} \sim N[(\mathbf{S}^{-1}\mathbf{A})^{\dagger}\mathbf{S}^{-1}\mathbf{A}\bar{\mathbf{x}}, (\mathbf{A}^T\mathbf{S}^{-2}\mathbf{A})^{\dagger}] \tag{6.4-38}$$

It follows directly, then, that the vector $\hat{\mathbf{x}} - (\mathbf{S}^{-1}\mathbf{A})^{\dagger}\mathbf{S}^{-1}\mathbf{A}\bar{\mathbf{x}}$ is normally distributed with mean $\mathbf{0}$ and variance matrix $(\mathbf{A}^T\mathbf{S}^{-2}\mathbf{A})^{\dagger}$; that is,

$$[\hat{\mathbf{x}} - (\mathbf{S}^{-1}\mathbf{A})^{\dagger}\mathbf{S}^{-1}\mathbf{A}\bar{\mathbf{x}}] \sim N[\mathbf{0}, (\mathbf{A}^T\mathbf{S}^{-2}\mathbf{A})^{\dagger}]. \tag{6.4-39}$$

The quadratic form $q(\hat{\mathbf{x}})$ defined by Eq. (6.4-36) can also be written

$$q(\hat{\mathbf{x}}) = [\hat{\mathbf{x}} - (\mathbf{S}^{-1}\mathbf{A})^{\dagger}\mathbf{S}^{-1}\mathbf{A}\bar{\mathbf{x}}]^T\mathbf{A}^T\mathbf{S}^{-2}\mathbf{A}[\hat{\mathbf{x}} - (\mathbf{S}^{-1}\mathbf{A})^{\dagger}\mathbf{S}^{-1}\mathbf{A}\bar{\mathbf{x}}]. \tag{6.4-40}$$

To see this, we expand the right-hand side of the foregoing equation, at the same time writing the factor $\mathbf{A}^T\mathbf{S}^{-2}\mathbf{A}$ in the form $\mathbf{A}^T\mathbf{S}^{-1}\mathbf{S}^{-1}\mathbf{A}$, and then use the identity (2.6-3) for generalized inverses to simplify the resulting expression. The result of this operation is

$$\begin{aligned}
&[\hat{\mathbf{x}} - (\mathbf{S}^{-1}\mathbf{A})^{\dagger}\mathbf{S}^{-1}\mathbf{A}\bar{\mathbf{x}}]^T\mathbf{A}^T\mathbf{S}^{-2}\mathbf{A}[\hat{\mathbf{x}} - (\mathbf{S}^{-1}\mathbf{A})^{\dagger}\mathbf{S}^{-1}\mathbf{A}\bar{\mathbf{x}}]\\
&\qquad = \hat{\mathbf{x}}^T\mathbf{A}^T\mathbf{S}^{-2}\mathbf{A}\hat{\mathbf{x}} - 2\bar{\mathbf{x}}^T\mathbf{A}^T\mathbf{S}^{-2}\mathbf{A}\hat{\mathbf{x}} + \bar{\mathbf{x}}^T\mathbf{A}^T\mathbf{S}^{-2}\mathbf{A}\bar{\mathbf{x}}\\
&\qquad = (\bar{\mathbf{x}} - \hat{\mathbf{x}})^T\mathbf{A}^T\mathbf{S}^{-2}\mathbf{A}(\bar{\mathbf{x}} - \hat{\mathbf{x}}),
\end{aligned}$$

which establishes (6.4.-40).

We can now generalize a theorem given by Anderson [3, Theorem 3.3.3, p. 54] to show that $q(\mathbf{x})$ is distributed according to a chi-square (χ^2) distribution with r degrees of freedom where r is the rank of the matrix $\mathbf{A}$. To do this, we first factor the symmetric matrix $\mathbf{A}^T\mathbf{S}^{-2}\mathbf{A}$ into the form

$$\mathbf{A}^T\mathbf{S}^{-2}\mathbf{A} = \mathbf{P}\boldsymbol{\Omega}^2\mathbf{P}^T \tag{6.4-41}$$

where $\mathbf{P}$ is an orthogonal matrix; that is,

$$\mathbf{P}^T\mathbf{P} = \mathbf{I}_n, \tag{6.4-42}$$

and $\boldsymbol{\Omega}^2$ is the diagonal matrix

$$\boldsymbol{\Omega}^2 = \operatorname{diag}(\omega_1^2, \omega_2^2, \ldots, \omega_r^2, \underbrace{0, \ldots, 0}_{n-r}) \tag{6.4-43}$$

whose diagonal elements are the eigenvalues of $\mathbf{A}^T\mathbf{S}^{-2}\mathbf{A}$. Then by Eq. (2.6-10), the generalized inverse of $\mathbf{A}^T\mathbf{S}^{-2}\mathbf{A}$ is given by

$$(\mathbf{A}^T\mathbf{S}^{-2}\mathbf{A})^{\dagger} = \mathbf{P}(\boldsymbol{\Omega}^2)^{\dagger}\mathbf{P}^T \tag{6.4-44}$$

where

$$(\boldsymbol{\Omega}^2)^{\dagger} = \operatorname{diag}\left(\frac{1}{\omega_1^2}, \frac{1}{\omega_2^2}, \ldots, \frac{1}{\omega_r^2}, \overbrace{0, \ldots, 0}^{n-r}\right). \tag{6.4-45}$$

From these two expressions, it follows that

$$\mathbf{P}^T(\mathbf{A}^T\mathbf{S}^{-2}\mathbf{A})^\dagger\mathbf{P} = \operatorname{diag}\left(\frac{1}{\omega_1^2}, \frac{1}{\omega_2^2}, \ldots, \frac{1}{\omega_r^2}, 0, \ldots, 0\right). \tag{6.4-46}$$

If $\mathbf{D}$ is the diagonal matrix defined by

$$\mathbf{D} = \operatorname{diag}(\omega_1, \omega_2, \ldots, \omega_r, \overbrace{1, \ldots, 1}^{n-r}) = \mathbf{D}^T, \tag{6.4-47}$$

then it is easy to see that

$$\mathbf{D}^T\mathbf{P}^T(\mathbf{A}^T\mathbf{S}^{-2}\mathbf{A})^\dagger\mathbf{P}\mathbf{D} = \operatorname{diag}(\overbrace{1, 1, \ldots, 1}^{r}, \overbrace{0, \ldots, 0}^{n-r}). \tag{6.4-48}$$

This expression can be written more simply as

$$\mathbf{C}(\mathbf{A}^T\mathbf{S}^{-2}\mathbf{A})^\dagger\mathbf{C}^T = \begin{pmatrix} \mathbf{I}_r & \mathbf{0} \\ \mathbf{0} & \mathbf{0} \end{pmatrix}, \tag{6.4-49}$$

where $\mathbf{C}$ is the nonsingular matrix

$$\mathbf{C} = \mathbf{D}^T\mathbf{P}^T \tag{6.4-50}$$

whose inverse can be written

$$\mathbf{C}^{-1} = \mathbf{P}\mathbf{D}^{-1} = \mathbf{P}\operatorname{diag}\left(\frac{1}{\omega_1}, \frac{1}{\omega_2}, \ldots, \frac{1}{\omega_r}, \overbrace{1, \ldots, 1}^{n-r}\right). \tag{6.4-51}$$

We now define a new random variable $\hat{\mathbf{z}}$ by

$$\hat{\mathbf{z}} = \mathbf{C}[\hat{\mathbf{x}} - (\mathbf{S}^{-1}\mathbf{A})^\dagger\mathbf{S}^{-1}\mathbf{A}\bar{\mathbf{x}}]. \tag{6.4-52}$$

Using the same theorem that was used to obtain the distribution of $\hat{\mathbf{x}}$ from $\hat{\mathbf{b}}$, it follows from expressions (6.4-39) and (6.4-52) that

$$\hat{\mathbf{z}} \sim N[\mathbf{0}, \mathbf{C}(\mathbf{A}^T\mathbf{S}^{-2}\mathbf{A})^\dagger\mathbf{C}^T],$$

which, by (6.4-49), can be written

$$\hat{\mathbf{z}} \sim N\left[\mathbf{0}, \begin{pmatrix} \mathbf{I}_r & \mathbf{0} \\ \mathbf{0} & \mathbf{0} \end{pmatrix}\right]. \tag{6.4-53}$$

This is a *singular* or *degenerate* normal distribution, in which all of the probability mass is concentrated in the r-dimensional subspace corresponding to the first r dimensions. The variables $\hat{z}_1, \hat{z}_2, \ldots, \hat{z}_r$ are independently normally distributed, each one with mean 0 and variance 1.

Solving Eq. (6.4-52) for $\hat{\mathbf{x}} - (\mathbf{S}^{-1}\mathbf{A})^\dagger\mathbf{S}^{-1}\mathbf{A}\bar{\mathbf{x}}$ gives

$$[\hat{\mathbf{x}} - (\mathbf{S}^{-1}\mathbf{A})^\dagger\mathbf{S}^{-1}\mathbf{A}\bar{\mathbf{x}}] = \mathbf{C}^{-1}\hat{\mathbf{z}},$$

and if this expression is substituted into Eq. (6.4-40), the result is

$$q(\hat{\mathbf{x}}) = \hat{\mathbf{z}}^T (\mathbf{C}^{-1})^T \mathbf{A}^T \mathbf{S}^{-2} \mathbf{A} \mathbf{C}^{-1} \hat{\mathbf{z}},$$

which, by Eqs. (6.4-51) and (6.4-41), can be written

$$q(\hat{\mathbf{x}}) = \hat{\mathbf{z}}^T \mathbf{D}^{-1} \mathbf{P}^T \mathbf{A}^T \mathbf{S}^{-2} \mathbf{A} \mathbf{P} \mathbf{D}^{-1} \hat{\mathbf{z}} = \hat{\mathbf{z}}^T \mathbf{D}^{-1} \mathbf{\Omega}^2 \mathbf{D}^{-1} \hat{\mathbf{z}}.$$

It is easy to see that

$$\mathbf{D}^{-1} \mathbf{\Omega}^2 \mathbf{D}^{-1} = \begin{pmatrix} \mathbf{I}_r & \mathbf{0} \\ \mathbf{0} & \mathbf{0} \end{pmatrix},$$

so that

$$q(\hat{\mathbf{x}}) = \hat{\mathbf{z}}^T \begin{pmatrix} \mathbf{I}_r & \mathbf{0} \\ \mathbf{0} & \mathbf{0} \end{pmatrix} \hat{\mathbf{z}} = \sum_{i=1}^{r} \hat{z}_i^2. \tag{6.4-54}$$

Combining the results (6.4-53) and (6.4-54), we have shown that the quadratic form

$$q(\hat{\mathbf{x}}) = (\bar{\mathbf{x}} - \hat{\mathbf{x}})^T \mathbf{A}^T \mathbf{S}^{-2} \mathbf{A} (\bar{\mathbf{x}} - \hat{\mathbf{x}})$$

can be expressed as the sum of the squares of r stochastically independent random variables $z_1, z_2, \ldots, z_r$, each of which is normally distributed with mean 0 and variance 1. From this fact, if follows [4, Theorem 1, p. 351] that $q(\hat{\mathbf{x}})$ is distributed according to a chi-square distribution with r degrees of freedom; that is,

$$q(\hat{\mathbf{x}}) = (\bar{\mathbf{x}} - \hat{\mathbf{x}})^T \mathbf{A}^T \mathbf{S}^{-2} \mathbf{A} (\bar{\mathbf{x}} - \hat{\mathbf{x}}) \sim \chi^2(r). \tag{6.4-55}$$

We can now relate the parameter γ^2, which determines the size of the confidence ellipsoid $E_\gamma(\hat{\mathbf{x}})$, to the probability level α associated with $E_\gamma(\hat{\mathbf{x}})$. The probability distribution function associated with the $\chi^2(r)$ distribution is given by

$$f(\rho) = \begin{cases} [\Gamma(\rho/2) 2^{r/2}]^{-1} \rho^{r/2-1} \exp(-\rho/2), & 0 \leqslant \rho < \infty, \\ 0, & \rho < 0, \end{cases} \tag{6.4-56}$$

where $\Gamma(\rho/2)$ denotes the gamma function, which is defined by

$$\Gamma(\eta) = \int_0^\infty y^{\eta-1} e^{-y} \, dy.$$

The probability that $q(\hat{\mathbf{x}})$ is less than or equal to the parameter γ^2 is

$$\Pr\{(\bar{\mathbf{x}} - \hat{\mathbf{x}})^T \mathbf{A}^T \mathbf{S}^{-2} \mathbf{A} (\bar{\mathbf{x}} - \hat{\mathbf{x}}) \leqslant \gamma^2\} = \int_0^{\gamma^2} f(\rho) \, d\rho,$$

so that for the confidence ellipsoid $E_\gamma(\hat{\mathbf{x}})$, we can write

$$\Pr\{\bar{\mathbf{x}} \in E_\gamma(\hat{\mathbf{x}})\} = \Pr\{(\bar{\mathbf{x}} - \hat{\mathbf{x}})^T \mathbf{A}^T \mathbf{S}^{-2} \mathbf{A} (\bar{\mathbf{x}} - \hat{\mathbf{x}}) \leqslant \gamma^2\} = \alpha \tag{6.4-57}$$

where

$$\alpha = \int_0^{\gamma^2} [\Gamma(\rho/2) 2^{r/2}]^{-1} \rho^{r/2-1} \exp(-\rho/2)\, d\rho. \qquad (6.4\text{-}58)$$

The values of γ^2 associated with various values of α and r are tabulated in Table 6.2.

TABLE 6.2 Values of γ^2 associated with various values of r and α

	α					
r	0.70	0.80	0.90	0.95	0.99	0.995
1	1.07	1.64	2.71	3.84	6.63	7.88
2	2.41	3.22	4.61	5.99	9.21	10.6
3	3.66	4.64	6.25	7.81	11.3	12.8
5	6.06	7.29	9.24	11.1	15.1	16.7
10	11.8	13.4	16.0	18.3	23.2	25.2
20	22.8	25.0	28.4	31.4	37.6	40.0
40	44.2	47.3	51.8	55.8	63.7	66.8
60	65.2	69.0	74.4	79.1	88.4	92.0

It is interesting to compare the size of the confidence ellipsoid $E_\gamma(\hat{\mathbf{x}})$ with that of a κ-ellipsoid associated with the same probability level α. Since the equations of the two ellipsoids are

$$(\mathbf{x} - \hat{\mathbf{x}})^T \mathbf{A}^T \mathbf{S}^{-2} \mathbf{A}(\mathbf{x} - \hat{\mathbf{x}}) = \gamma^2,$$

$$(\mathbf{x} - \hat{\mathbf{x}})^T \mathbf{A}^T \mathbf{S}^{-2} \mathbf{A}(\mathbf{x} - \hat{\mathbf{x}}) = \kappa^2;$$

the ratio γ/κ is the factor by which the κ-ellipsoid must be expanded in order to make it identical with the confidence ellipsoid. This ratio is tabulated for several values of α and r in Table 6.3. For each α, as the value of r increases, the expansion factor increases also, but higher values of r correspond to higher-dimensional problems; and in such problems, the nonnegativity constraint becomes more and more important in limiting the size of the constraint region for the problem, since, in general, a smaller and smaller percentage of the confidence ellipsoid will lie in the positive orthant. Furthermore, for higher probability levels α, the rate at which the expansion factor increases with increasing values of r is less than the rate for lower values of α.

Now that we have obtained the relationship between γ^2 and α for the confidence ellipsoid, we must find expressions for the end points of the confidence interval $I_\gamma^E(\phi)$ defined by Eq. (6.4-35). The quantities ϕ_E^{lo} and ϕ_E^{up} are the minimum and maximum values of the function $\phi(\mathbf{x}) = \mathbf{c}^T\mathbf{x}$ when $\mathbf{x}$ is constrained to

TABLE 6.3

	α: 0.70	0.90	0.95	0.99
	κ: 1.04	1.64	1.96	2.58
$r=1$, γ	1.04	1.64	1.96	2.58
γ/κ	1.0	1.0	1.0	1.0
$r=2$, γ	1.55	2.15	2.45	3.03
γ/κ	1.49	1.31	1.25	1.17
$r=5$, γ	2.46	3.04	3.33	3.89
γ/κ	2.36	1.85	1.70	1.51
$r=10$, γ	3.44	4.00	4.28	4.82
γ/κ	3.31	2.44	2.18	1.87
$r=20$, γ	4.77	5.33	5.60	6.13
γ/κ	4.58	3.25	2.86	2.38
$r=40$, γ	6.65	7.20	7.47	7.98
γ/κ	6.40	4.39	3.81	3.09
$r=60$, γ	8.07	8.63	8.89	9.40
γ/κ	7.76	5.26	4.53	3.64

lie in the confidence ellipsoid $E_\gamma(\hat{\mathbf{x}})$ and at the same time required to lie in the positive orthant. Therefore,

$$\phi_E^{\mathrm{lo}} = \min_{\mathbf{x}\geqslant\mathbf{o}} \{\mathbf{c}^T\mathbf{x} \mid (\mathbf{x}-\hat{\mathbf{x}})^T\mathbf{A}^T\mathbf{S}^{-2}\mathbf{A}(\mathbf{x}-\hat{\mathbf{x}})^T \leqslant \gamma^2\},$$

$$\phi_E^{\mathrm{up}} = \max_{\mathbf{x}\geqslant\mathbf{o}} \{\mathbf{c}^T\mathbf{x} \mid (\mathbf{x}-\hat{\mathbf{x}})^T\mathbf{A}^T\mathbf{S}^{-2}\mathbf{A}(\mathbf{x}-\hat{\mathbf{x}}) \leqslant \gamma^2\},$$

or

$$\phi_E^{\mathrm{lo}} = \min_{\mathbf{x}\geqslant\mathbf{o}} \{\mathbf{c}^T\mathbf{x} \mid (\hat{\mathbf{b}}-\mathbf{A}\mathbf{x})^T\mathbf{S}^{-2}(\hat{\mathbf{b}}-\mathbf{A}\mathbf{x}) \leqslant r_0+\gamma^2\},$$

$$\phi_E^{\mathrm{up}} = \max_{\mathbf{x}\geqslant\mathbf{o}} \{\mathbf{c}^T\mathbf{x} \mid (\hat{\mathbf{b}}-\mathbf{A}\mathbf{x})^T\mathbf{S}^{-2}(\hat{\mathbf{b}}-\mathbf{A}\mathbf{x}) \leqslant r_0+\gamma^2\},$$

where r_0 is minimum value of the quadratic form $(\hat{\mathbf{b}}-\mathbf{A}\mathbf{x})^T\mathbf{S}^{-2}(\hat{\mathbf{b}}-\mathbf{A}\mathbf{x})$ [r_0 is given by Eq. (6.4-19)]. Using norm notation, these problems can be written

$$\phi_E^{\mathrm{lo}} = \min_{\mathbf{x}\geqslant\mathbf{o}} \{\mathbf{c}^T\mathbf{x} \mid \|\mathbf{A}\mathbf{x}-\hat{\mathbf{b}}\|_{\mathbf{S}^2}^2 \leqslant r_0+\gamma^2\},$$

$$\phi_E^{\mathrm{up}} = \max_{\mathbf{x}\geqslant\mathbf{o}} \{\mathbf{c}^T\mathbf{x} \mid \|\mathbf{A}\mathbf{x}-\hat{\mathbf{b}}\|_{\mathbf{S}^2}^2 \leqslant r_0+\gamma^2\}.$$

The quantities ϕ_E^{lo} and ϕ_E^{up} can be evaluated by the parametric quadratic programming technique described in Section 5.4.

In Section 6.1, we discussed the use of constrained estimation techniques in solving estimation problems involving systems of Fredholm equations of the form

$$\int_a^b K_i(s)\,x(s)\,ds = \hat{y}_i - \hat{e}_i, \qquad i = 1, 2, \ldots, m, \tag{6.4-59}$$

where $x(s)$ is an unknown function, the $K_i(s)$ are known functions, the $\hat{y}_i$ are known ("measured") values, and the $\hat{e}_i$ are errors associated with the $\hat{y}_i$. The error vector $\hat{\mathbf{e}}$ was assumed to be a random sample from one of the fixed, bounded regions defined by Eqs. (6.1-3), (6.1-4), and (6.1-5). The quantity to be estimated had the form

$$\phi = \int_a^b w(s)\,x(s)\,ds \tag{6.4-60}$$

where $w(s)$ is a known function. The interval estimates $[\phi^{lo}, \phi^{up}]$ that were obtained were guaranteed to contain the true value of ϕ because the error vector $\hat{\mathbf{e}}$ was chosen with 100% probability from a finite bounded region. It should be evident from the preceding discussion in this section that the problem above can easily be generalized to the statistical problem in which the $\hat{\mathbf{e}}$ vector is a sample from a multivariate normal distribution. This is a problem that arises very often in the physical sciences. The discretization of the problem is analogous to that for the nonstochastic problem defined by error distribution (6.1-4), but the interval estimates obtained will be confidence interval estimates of the form $[\phi^{lo}(\alpha), \phi^{up}(\alpha)]$ where α is the parameter determining the level of confidence. The arguments relating this level of confidence to the interval obtained are analogous to the arguments given earlier in this section for the discrete vector-matrix problem.

In solving for confidence intervals using the constrained estimation technique, it sometimes happens that for a given confidence level α, a confidence interval does not exist. This situation arises when the ordinary $100\alpha\%$ confidence ellipsoid does not have any points in common with the positive orthant. There are two ways in which this situation may be remedied. One is simply to increase the confidence level α, thus expanding the size of the $100\alpha\%$ confidence ellipsoid until it has a nonzero intersection with the positive orthant. This technique is illustrated in Fig. 6.12a for a problem in which $r = n = 2$ and $\mathbf{c} = \mathbf{e}_1 = (1, 0)^T$. The $100\alpha\%$ confidence ellipsoid does not give a valid confidence interval for the constrained estimation problem because it does not have an intersection with the positive orthant. But the $100\alpha'\%$ confidence ellipsoid does have a nonzero intersection with the positive orthant. It is shown as a shaded region. The support planes of this shaded region, which are orthogonal to c, would give the $100\alpha'\%$ confidence interval.

The other method for obtaining a confidence ellipsoid is to replace the constraint

$$(\hat{\mathbf{b}} - \mathbf{K}\mathbf{x})^T \mathbf{S}^{-2}(\hat{\mathbf{b}} - \mathbf{K}\mathbf{x}) \leq r_0 + \kappa^2$$

by the weaker constraint

$$(\hat{\mathbf{b}} - \mathbf{K}\mathbf{x})^T \mathbf{S}^{-2}(\hat{\mathbf{b}} - \mathbf{K}\mathbf{x}) \leq r_0' + \kappa^2 \tag{6.4-61}$$

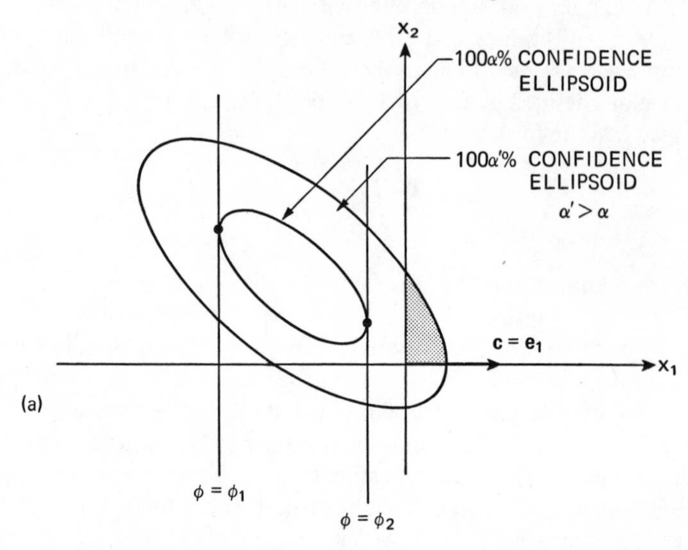

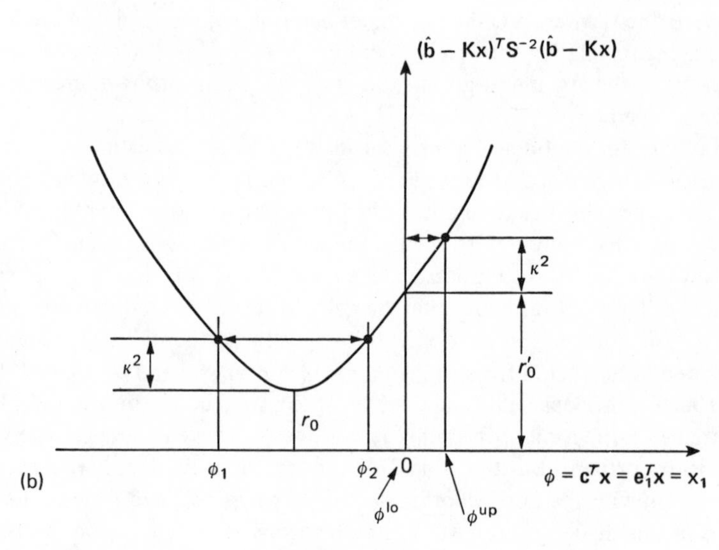

Figure 6.12.

where

$$r_0' = \min_{\mathbf{x} \geq \mathbf{0}} \{(\hat{\mathbf{b}} - \mathbf{Kx})^T \mathbf{S}^{-2} (\hat{\mathbf{b}} - \mathbf{Kx})\} \geq r_0. \qquad (6.4\text{-}62)$$

This situation is illustrated in Fig. 6.12b for the same problem that was shown in Fig. 6.12a. The interval $[\phi^{\mathrm{lo}}, \phi^{\mathrm{up}}]$ is taken as the $100\alpha\%$ confidence level

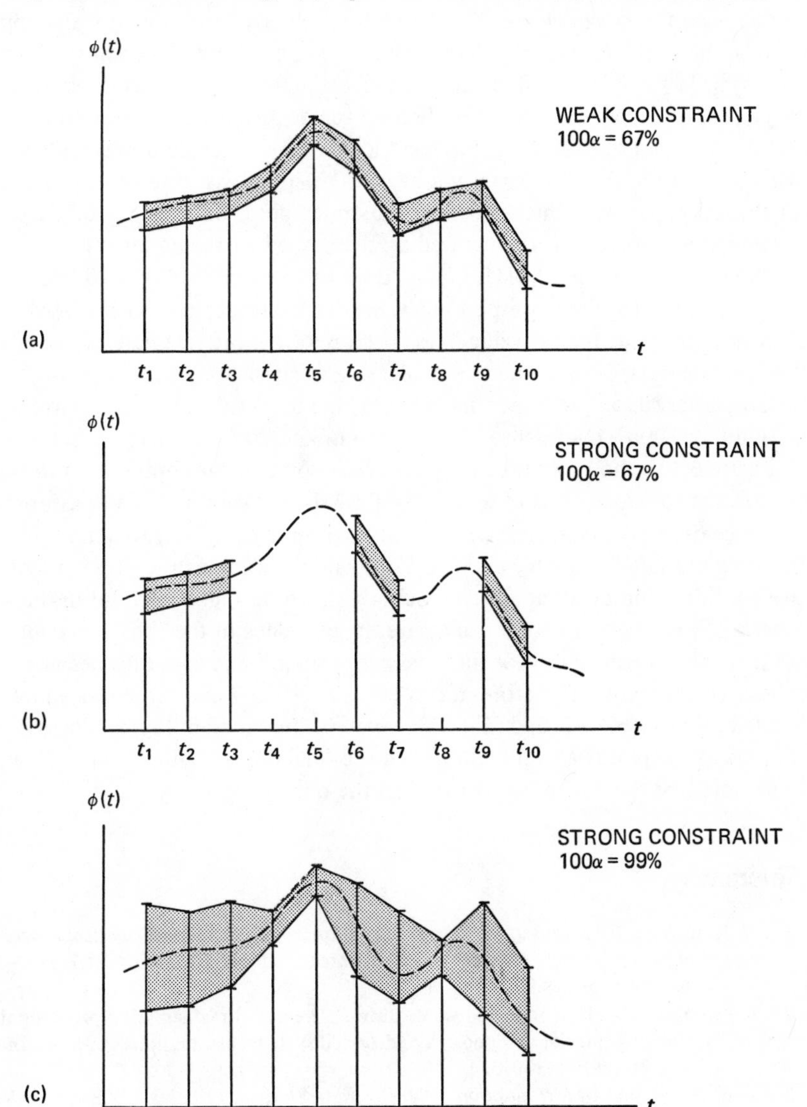

Figure 6.13.

even though a higher level of confidence could be claimed for it. Thus, we have

$$\Pr\{\phi^{\mathrm{lo}} \leqslant \phi \leqslant \phi^{\mathrm{up}}\} \geqslant \alpha \tag{6.4-63}$$

rather than

$$\Pr\{\phi^{\mathrm{lo}} \leqslant \phi \leqslant \phi^{\mathrm{up}}\} = \alpha$$

in this case. The advantage of this procedure is chiefly aesthetic. In many applications, the problem is to calculate a whole series of interval estimates $[\phi_1^{\mathrm{lo}}, \phi_1^{\mathrm{up}}]$, $[\phi_2^{\mathrm{lo}}, \phi_2^{\mathrm{up}}]$, $[\phi_3^{\mathrm{lo}}, \phi_3^{\mathrm{up}}]$, . . ., which are interval estimates for the quantities $\phi(t_1)$, $\phi(t_2), \phi(t_3)$, . . ., which in turn are discrete values of some continuous function $\phi(t)$. It is desired to obtain a $100\alpha\%$ confidence level estimate for each of these discrete values so that the result will be a $100\alpha\%$ pointwise interval approximation for the unknown $\phi(t)$. The situation is illustrated graphically in Fig. 6.13a. The unknown function $\phi(t)$ is represented by the dashed curve and interval estimates are shown for ten points $t_1, t_2, \ldots, t_{10}$. Each of these intervals is a $100\alpha\%$ confidence interval for the corresponding value $\phi(t_i)$. Although no such probabilistic claims can be made for the value of $\phi(t)$ at any of the intermediate values of t, it is, nevertheless, often aesthetically pleasing to connect up the endpoints of all these intervals and to hope that the resulting band forms a crude approximation to a $100\alpha\%$ confidence band for the unknown function.

Figure 6.13b shows what can happen when some of the confidence intervals fail to exist for some value of α, say, $\alpha = 0.67$. It is always possible to assure the existence of the confidence interval at all the points $t_1, t_2, \ldots, t_{10}$ simply by choosing the value of α large enough. For example, a value of $\alpha = 0.99$ might give confidence intervals at all ten points as shown in Fig. 6.13c. But in the process of increasing α in order to assure the existence of the confidence interval at t_4, t_5, and t_8, the widths of the intervals at the other values of t_i became unaesthetically large. Therefore, if α is taken to be 0.67 and if the procedure defined by Eqs. (6.4-61) and (6.4-62), which is illustrated in Fig. 6.12b, is adopted at the points t_4, t_5, and t_8, then the final 67% "confidence band" will look more like the one in Fig. 6.13a than the one in Fig. 6.13c.

References

1 W. R. Burrus, *Utilization of A Priori Information by Means of Mathematical Programming in the Statistical Interpretation of Measured Distributions,* ORNL-3743, June, 1965, pp. 49–50.

2 J. Replogle, B. D. Holcomb, and W. R. Burrus, The use of mathematical programming for solving singular and poorly conditioned systems of equations, *J. Math. Anal. Appl.,* **20** (1967), 310–324.

3 T. W. Anderson, *An Introduction to Multivariate Statistical Analysis,* Wiley, New York, 1958.

4 R. V. Hogg and A. T. Craig, *Introduction to Mathematical Statistics,* Macmillan, New York, 1965.

Author Index

Page numbers set in italics designate those pages on which the complete literature citation is given.

Numbers in parentheses designate the reference numbers where information is given.

Subject Index